T0324674

GRAVITATIONAL LENSING

Gravitational lensing is a consequence of general relativity, where the gravitational force due to a massive object bends the paths of light originating from distant objects lying behind it. Using very little general relativity and no higher level mathematics, this text presents the basics of gravitational lensing, focusing on the equations needed to understand the phenomenon. It then applies them to a diverse set of topics, including multiply imaged objects, time delays, extrasolar planets, microlensing, cluster masses, galaxy shape measurements, cosmic shear, and lensing of the cosmic microwave background. This approach allows undergraduate students and others to get quickly up to speed on the basics and the important issues. The text will be especially relevant as large surveys such as LSST and Euclid begin to dominate the astronomical landscape. Designed for a one-semester course, it is accessible to anyone with two years of undergraduate physics background.

SCOTT DODELSON is a Distinguished Scientist at Fermilab and Professor in the Department of Astrophysics and Astronomy at the University of Chicago. He has written more than 180 research papers on the connections between physics and astronomy. He is the author of *Modern Cosmology,* a standard graduate textbook, and he has recently taken leadership roles in surveys that employ gravitational lensing as a tool to get at basic physics. Through these roles and by teaching courses devoted to lensing, he realized that learning enough about lensing to begin research is both simple and rewarding. This volume was written to help potential researchers at all levels learn this fascinating, rapidly emerging field.

GRAVITATIONAL LENSING

SCOTT DODELSON

Fermi National Accelerator Laboratory, Batavia, Illinois

CAMBRIDGE
UNIVERSITY PRESS

CAMBRIDGE
UNIVERSITY PRESS

Shaftesbury Road, Cambridge CB2 8EA, United Kingdom

One Liberty Plaza, 20th Floor, New York, NY 10006, USA

477 Williamstown Road, Port Melbourne, VIC 3207, Australia

314–321, 3rd Floor, Plot 3, Splendor Forum, Jasola District Centre, New Delhi – 110025, India

103 Penang Road, #05–06/07, Visioncrest Commercial, Singapore 238467

Cambridge University Press is part of Cambridge University Press & Assessment,
a department of the University of Cambridge.

We share the University's mission to contribute to society through the pursuit of
education, learning and research at the highest international levels of excellence.

www.cambridge.org
Information on this title: www.cambridge.org/9781107129764

© Scott Dodelson 2017

This publication is in copyright. Subject to statutory exception and to the provisions
of relevant collective licensing agreements, no reproduction of any part may take
place without the written permission of Cambridge University Press & Assessment.

First published 2017

A catalogue record for this publication is available from the British Library

ISBN   978-1-107-12976-4   Hardback

Additional resources for this publication at www.cambridge.org/9781107129764

Cambridge University Press & Assessment has no responsibility for the persistence
or accuracy of URLs for external or third-party internet websites referred to in this
publication and does not guarantee that any content on such websites is, or will
remain, accurate or appropriate.

This book is dedicated to the memory of Danielle Bessler, who saw life through her own lens, thereby bringing joy to her family and friends

Contents

Color plates section can be found between pages 116 and 117

Preface

There are three reasons why I wanted to write this book. First, gravitational lensing is emerging as a powerful tool in many areas of astronomy, from exoplanets to cosmology. This breadth of applications emerging from a single phenomenon – the bending of light by curved space-time – is the second reason, for it is yet another example of why physics is so interesting: we can explain many things with a few simple laws.

The final reason for this book is that, while important for many different applications, lensing is not that hard. There have been elegant, beautiful papers and books exploring the foundations of lensing and many of its formal aspects. But most of the applications simply do not require all that much formalism. General relativity makes several appearances here, but even if you skip those few sections, you will still learn essentially everything that's here: multiple images, magnification, micro-lensing, shear, etc. And armed with that information, I hope you will be able to begin research on real problems that are opening up in so many areas of astronomy and be prepared to analyze the ever-improving datasets that are coming our way.

The goal then is that this book can serve as the text for an undergraduate or graduate course on gravitational lensing or be used for independent study by someone interested in jumping into research.

Thanks are due to colleagues who generously gave of their time to answer questions: Gary Bernstein, Daniel Fabrycky, Bhuvnesh Jain, Rachel Mandelbaum, and Ben Wandelt. I am very grateful to the students who took this course, especially in 2015, as they helped immensely by providing feedback on an early draft of this book. So thank you to these rising stars: Adam Anderson, Gourav Khullar, Meng-Xiang Lin, Monica Mocanu, Pavel Motloch, Andrew Neil, Zhaodi Pan, Jason Poh, Amy Tang, Rito Thakur, and Alexander Tolish.

I am extraordinarily grateful to the people who fund my research, especially Fermilab, the Office of High Energy Physics at the Department of Energy and, by extension, the citizens of the United States. I am so fortunate to live in a society that values basic research. Some of this work was carried out at the Aspen Center for Physics, which is supported by National Science Foundation grant PHY-1066293.

1

Overview

Light passing by a point particle with mass M is deflected (Fig. 1.1) by an angle

$$\delta\theta = \frac{4MG}{bc^2} \tag{1.1}$$

where b is the distance of closest approach, the *impact parameter*; G is Newton's constant; and c is the speed of light.

Physically and dimensionally, this makes sense: the deflection angle is bound to increase as the mass of the deflector, or *lens*, gets larger and decrease as the light path strays farther from the lens. And, of course, the deflection is caused by gravity, so must be proportional to Newton's constant.

The factor of 4 is trickier and carries an interesting history. When developing his theory of gravity, general relativity, Einstein initially derived a result with coefficient equal to 2, a prediction that agreed with one obtained a century earlier that relied on Newtonian physics. Based on this result, he calculated the deflection angle of a light ray coming from a star aligned with the edge of the Sun and determined it to be a little less than an arcsecond. Einstein sent eminent astronomer George Hale a letter asking if a deviation of this size could be detected. Hale responded that the starlight would be dominated by light from the Sun unless the Sun were eclipsed. So, Einstein courted the German astronomer Erwin Freundlich to undertake an expedition to the site at which the next eclipse would be maximal: the Crimean Peninsula in Summer 1914. It turns out that August 1914 was not a very good time to be wandering around the continent, and Freundlich and his team were imprisoned. The good news is that they were released eventually; the apparent bad news is that they were released after the eclipse was over, so could not make the observations that Einstein deemed "a simply invaluable service to theoretical physics." This was only *apparent* bad news, because Einstein subsequently tweaked his theory and – as we will see – got a different coefficient for the deflection.

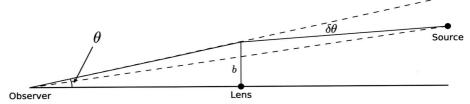

Figure 1.1 Light from a source (far right) is deflected by an angle $\delta\theta$ as it passes a projected distance b by a lens of mass M. The source appears to the observer to be an angular distance θ away from the line of sight connecting the observer to the lens.

Figure 1.2 Photograph of the 1919 solar eclipse by Dyson et al. (1920).

In May 1919, Sir Frank Watson Dyson and Sir Arthur Eddington led an expedition to the island of Príncipe, off the west coast of Africa, and sent another team to Brazil in case the weather was cloudy. They were to observe the positions of stars in the Hyades cluster – which was situated right near the limb of the Sun – to see if they were shifted by the mass of the Sun. Their team took the photograph shown in Fig. 1.2. The positions of the background stars were indeed off from their expected positions by 1.9″, in agreement with the (improved) theory. Einstein became a celebrity and the theory of general relativity officially became the theory of gravity.

The success of the 1919 expeditions is often heralded as a crowning achievement, in that it secured Einstein's fame and general relativity's place in the pantheon of fundamental physics laws. This is true, but this "game-ending" interpretation casts a shadow over perhaps an even more important ramification. The expeditions and the success of general relativity paved the way for a new field of astrophysics: gravitational lensing.

The path from 1919 to what might be called the first direct detection of a lensing event by Walsh et al. (1979) was so long in large part because instrumentation needed to catch up with theory. But there is another reason for the delay: since we usually do not know the true position of astronomical objects, it is difficult to understand how or when the deflection of light might be detected. If we don't know where things really are, then how can we tell that they do not appear where they should be? The case of the starlight passing by the Sun appears to be an almost unique case, where we know the actual location of the source and the lens and the mass of the lens. In general we will not have these advantages, so what are the observable effects of light deflection? The rest of this chapter provides a first glimpse at the answers to this question. The phenomena outlined here and explored in detail in the book have enabled gravitational lensing to play a powerful role in many areas of modern astronomy and cosmology.

Lensing phenomena vary in their complexity, ranging from the simple case of a point source lensed by a point mass (leftmost in Fig. 1.3) to a diffuse source lensed by a diffuse mass distribution (lower right). The outline in this chapter and the details in the ensuing chapters move gradually from the simple to the complex, with the aim of unifying these disparate phenomena and driving home the point that they all emerge from the same relatively simple physical law.

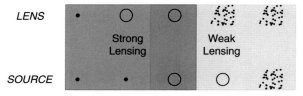

Figure 1.3 Cartoon depicting the range of phenomena spanned by gravitational lensing. The leftmost limit is when a point mass lenses a point source. Moving to the right, first the lens and then both the source and lens can be extended but are still single objects. The effects seen in these cases – multiple images, magnification, and microlensing – are associated with *strong lensing*. Moving further to the right, a single extended source can be distorted by an intervening diffuse set of lenses, or equivalently a fluctuating gravitational potential. The most extreme example of lensing is when the source itself is diffuse, for example, the cosmic microwave background. These less dramatic phenomena are detectable only statistically and fall in the domain of *weak lensing*.

1.1 Multiple Images

We begin our overview of lensing with the simple case of a point source lensed by a point mass, so that the light from the source is deflected by the amount given in Eq. (1.1). If the deflection angle is much smaller than the separation of the source from the lens-observer line of sight, there will be little observable effect. The more dramatic effects of lensing occur for objects with separation θ smaller than the deflection angle:

$$\theta < \delta\theta. \tag{1.2}$$

If the distance between the lens and us is denoted D_L, then the transverse distance b that represents the distance of closest approach corresponds to an angular size $\theta = b/D_L$[1] away from the line of sight. Therefore, the above requirement becomes

$$\theta < \frac{4MG}{D_L \theta c^2} \tag{1.3}$$

or

$$\theta < \sqrt{\frac{4MG}{D_L c^2}} = \theta_E, \tag{1.4}$$

where θ_E is called the Einstein radius. This is a good enough estimate for our purposes now, but in Chapter 2, we will derive a slightly modified expression for the Einstein radius of a point mass that accounts for the finite distance between us and the source. More generally, extended mass distributions have different coefficients, but Eq. (1.4) gives a good sense of when interesting effects will occur.

As shown in Fig. 1.4, if the source is within the Einstein radius of the lens, then multiple images can be observed. Most dramatic of all, if the source is directly behind the lens, then, as seen in Figure 1.5, the image will be a ring around the lens, a so-called *Einstein ring*, with radius equal to the Einstein radius.

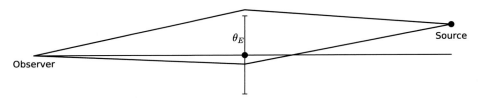

Figure 1.4 Light rays emanating in different directions from a source can be focused by an intervening lens on to the same point beyond the lens. The observer at this point will detect multiple images of the same object.

---

[1] Astronomy almost always works in the small angle approximation wherein $\sin\theta \simeq \tan\theta \simeq \theta$.

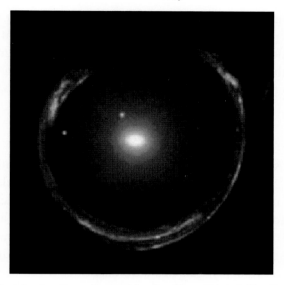

Figure 1.5 Einstein ring first observed by the Sloan Digital Sky Survey and then followed up with the Hubble Space Telescope. Foreground lens is a galaxy at the center, while the background object is almost perfectly aligned so is seen as a ring (Credit: ESA/Hubble & NASA). (See color plates section.)

Let's estimate a typical value for the Einstein radius. Normalize to a solar mass lens a distance of 1 kpc from us, so that

$$\theta_E = 1.4 \times 10^{-8} \left(\frac{\text{kpc}}{D_L}\right)^{1/2} \left(\frac{M}{M_\odot}\right)^{1/2}. \tag{1.5}$$

So every object has an Einstein radius, which increases with its mass and decreases the farther away it is from us. Translating to arcseconds, this becomes

$$\theta_E = 0.0028'' \left(\frac{\text{kpc}}{D_L}\right)^{1/2} \left(\frac{M}{M_\odot}\right)^{1/2}. \tag{1.6}$$

Typical angular resolutions for optical telescopes are in the arcsecond range, so a star in the Milky Way does lens distant objects but not in any way that can produce observable multiple images. At cosmological distances $D_L \sim 10^6$ kpc, a galaxy with mass of $10^{12} M_\odot$ has an Einstein radius of order an arc second, so can potentially produce multiple images of a distant point source.

1.2 Time Delay

Multiple images and most other lensing phenomena stem from the perpendicular deflection of light as it traverses past masses, or equivalently through a varying gravitational potential. But gravity causes not just changes in the perpendicular direction of the photons but also distortions in the propagation along the line of sight, i.e., time delays. Roughly, the time delay is of order the gravitational potential divided by the speed of light squared (MG/Rc^2 for a point mass) multiplied by the time spent in the neighborhood of the potential, R/c. So, we expect a time delay of order

$$\delta t \sim \frac{MG_\odot}{c^3} \sim 5\,\mu\text{sec} \tag{1.7}$$

for a light ray passing by the Sun. There are various factors of order unity that push this up to about 200 μsec, an effect dubbed the Shapiro time delay.

In the case of multiple images, the different light rays that reach us take different paths and therefore arrive at different times. If the source has a time dependence (i.e., is variable), the time delay can be observed, thereby obtaining information about both the lens and the various distances. This has the potential to be a powerful cosmological tool. Roughly, galaxies have masses of order $10^{12} M_\odot$ so delays of order $\delta t \sim 5 \times 10^6$ sec, or a few months, are expected and have been observed.

1.3 Magnification

Even if multiple images cannot be resolved, we might still observe the effect in the form of magnification since we receive flux from multiple directions. More generally, lensing conserves surface brightness (Exercise (1.3)), so the flux from an object depends on its apparent size. The magnification is given by the ratio of the lensed to unlensed sizes, as depicted in Fig. 1.6. For sources whose light passes within an Einstein radius of the lens ($\theta < \theta_E$) and whose true distance β from the line of sight connecting the observer to the lens is smaller than the Einstein radius ($\beta < \theta_E$), the magnification is of order

$$\mu \simeq \frac{\theta_E}{\beta}. \tag{1.8}$$

Therefore, even if multiple images cannot be observed by resolving them, they might contribute to an observed magnification of an object.

1.4 Microlensing

The discussion of magnification leads to the conclusion that a moving object passing between us and a distant star magnifies the star roughly for as long as the object is within the Einstein radius of the line of sight connecting us to the star. This short-term magnification is called *microlensing*. To get an estimate for the timescales, an object moving perpendicular to the line of sight with velocity v spends a time

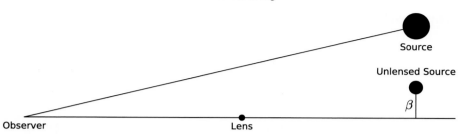

Figure 1.6 Magnification of an image is due to the increased area of a source. If the unlensed distance between the source and the line of sight connecting the observer to the lens β is much smaller than the Einstein radius, then this magnification is of order θ_E/β.

$$t \sim \frac{D_L \theta_E}{v} \tag{1.9}$$

within the Einstein radius of the line of sight. So the flux from the background star will be magnified for that brief amount of time. Let's plug in typical numbers for a Galactic lens: $D_L \sim 10$ kpc, $v \sim 200$ km/sec, and $M \sim M_\odot$. Then, the magnification will occur for a time

$$t \sim 6.8 \times 10^6 \text{ sec } \left(\frac{D_L}{10\,\text{kpc}}\right)^{1/2} \left(\frac{M}{M_\odot}\right)^{1/2}, \tag{1.10}$$

a few months.

Although the lens could be a star, a much more intriguing possibility is that the lens is something that could not be seen directly. For example, Fig. 5.5 shows an example of a background star lensed by a dark foreground object. A lens like this in our Galaxy is called a MAssive Compact Halo Object (MACHO). MACHOs were proposed as a candidate for the dark matter in the Galaxy, with the name a lighthearted reference to an alternative candidate, Weakly Interacting Massive Particles (WIMPs). We have observed many such events – Fig. 5.5 shows one – and an important question that has apparently been resolved in the negative is whether MACHOs might be the sole component of dark matter in the Galaxy.

If we have detected a background star lensed by a moving foreground object in the manner described above, we can hunt for planets revolving around the foreground star. The effect is more pronounced than one might imagine: the planet does not simply add to the mass of the host star when it is aligned; rather, it leads to a significant bump in the magnification for the short period of time that the aligned system is within the Einstein radius of the planet. A Jupiter-sized planet has an Einstein radius 30 times smaller than the Sun (since $\theta_E \propto M^{1/2}$) so, from Eq. (1.10), the blip it produces in the light curve (as shown in Fig. 5.11) will last for a period of time on the order of a day. An Earth-sized planet will produce a signal shorter by another factor of $1000^{1/2}$, or roughly an hour. Astronomers have monitored lensed stars in this way and derived the remarkable fact that there are more planets than stars in the Galaxy.

1.5 Extended Lenses

The first way to generalize beyond point masses is to envision an extended lens (but still a single lens), characterized not by a mass M but by a 2D surface density (Exercise (1.4)) $\Sigma(R)$. The mass enclosed within a radius R is then roughly $\pi \Sigma(R) R^2$, so the generalization of Eq. (1.4) would then be

$$\theta_E \simeq \sqrt{\frac{4\pi G \Sigma R^2}{D_L c^2}} = \sqrt{\frac{4\pi G \Sigma D_L}{c^2}}\, \theta. \tag{1.11}$$

We can rewrite this as

$$\left(\frac{\theta_E}{\theta}\right)^2 = \frac{\Sigma}{\Sigma_{\text{cr}}}, \tag{1.12}$$

where the critical surface density is approximately[2]

$$\Sigma_{\text{cr}} \simeq \frac{c^2}{4\pi D_L G}. \tag{1.13}$$

So light passing by regions in which the density is greater than the critical surface density will be within the Einstein radius of the lens and will therefore produce the types of dramatic changes described above. Plugging in the cosmological distance $D_L \simeq 1$ Gpc leads to a cosmological value of $\Sigma_{\text{cr}} \simeq 2 \times 10^{13} M_\odot/(100\text{kpc})^2$. So galaxy clusters, which are roughly this massive and above, can dramatically impact images that lie close to their centers. An example is shown in Fig. 1.7.

1.6 Extended Sources

The distortions evident in Fig. 1.7, background galaxies that appear highly elongated, provide a segue to the next level of generalization: the case where not only the lens but the source as well is extended. When an individual object serves as a lens for background galaxies, the resulting pattern is characteristic: ellipticities oriented tangentially to the line of sight connecting them to the lens center. This *tangential shear*, we will see, is directly related to the surface density; roughly,

$$\gamma_t(\theta) \sim \frac{\Sigma(\theta)}{\Sigma_{\text{cr}}}. \tag{1.14}$$

---

[2] The exact expression will be derived in Chapter 2 and includes a ratio of distances.

Figure 1.7 The shapes of background galaxies made highly elliptical by the fore-ground Abell cluster (Credit: Gravitational lensing in galaxy cluster Abell 2218: NASA, A. Fruchter and the ERO Team, STScI). (See color plates section.)

It is worth walking through the reason why the background galaxies in Fig. 1.7 are so elongated. A simple way to understand this is to remember that a source directly behind a lens leads to an Einstein ring, which is pretty close to what is observed in Fig. 1.7. Another way to think about why lenses make background objects more elliptical is to look at the cartoon in Fig. 1.8.

We are now in the middle of our progression across Fig. 1.3, with an extended source and lens. This is the region where both strong and weak lensing are impor-tant. Dramatic effects, characteristic of *strong lensing*, occur when $\Sigma > \Sigma_{cr}$, while small distortions, or *weak lensing*, occur in the low surface density limit.

It is illuminating to understand the reason why this case, single extended source and lens, allows for both strong and weak lensing. As mentioned above, the critical surface density for an object at a cosmological distance is of order $\Sigma_{cr} = 2 \times 10^{13} M_{\odot}/(100\text{kpc})^2$. As a simple example, let us assume that the density profile of a galaxy cluster is isothermal, so that

$$\rho(r) = \frac{\sigma^2}{2\pi G r^2},$$
(1.15)

where σ is the velocity dispersion, which is typically of order 1000 km/sec. The surface density of this cluster is then (Exercise (1.5)):

Figure 1.8 Background galaxies appear elliptical due to the deflection by a foreground lens. Left panel shows background circular galaxy with mass clump denoted by dots between us and the background galaxy. Middle panel shows that the light that passes closest to the mass will be deflected most, so the ensuing image we observe will look like the right panel: elliptical.

$$\Sigma(R) = \frac{\sigma^2}{2RG}. \qquad (1.16)$$

So $\Sigma(R)/\Sigma_{cr} = 2\pi D_L(\sigma/c)^2/R$. Since $\sigma/c \sim 1/300$, this ratio is unity when $R \simeq 2\pi \times 10^{-5}$ Gpc or 60 kpc. For a cluster with radius of order a Mpc, then, only a small fraction of the total area will be in the region where $\Sigma(R) > \Sigma_{cr}$; since area scales as R^2, that fraction will be $(60/1000)^2 \sim 0.004$. That means that, of all the background galaxies within a projected radius of 1 Mpc from the lens, fewer than one percent will be strongly lensed. All the others will be only weakly distorted by the foreground cluster. So, yes, there will be strong lensing phenomena, as is apparent in Fig. 1.7, but there will also be many, many galaxies in the weak lensing regime.

Another way of looking at this is to compute the total number of background galaxies that might be observed in the strong lensing regime. Taking the radius within which strong lensing occurs to be 60 kpc and a typical distance to a cluster to be 1000 Mpc, then the angular radius is a little more than $10''$. There are simply not that many galaxies in a region of this size. A final look at this is to think about adding the signal from every background galaxy, the signal being the tangential shear γ_t. The signal will typically be small compared to the noise, so the way to win is to beat down the noise: the more galaxies you measure, the smaller the noise becomes. We will encounter this often throughout the text, with the reduction in noise scaling as the square root of the number of objects sampled.

In this case, then, consider the signal-to-noise from a single annulus a distance R from the cluster center. In that annulus, the total number of galaxies is proportional to $R\,dR$, where dR is the width of the annulus. The signal-to-noise therefore scales as $R(dR/R)^{1/2}\gamma_t \propto (d\ln R)^{1/2}$, since $\gamma_t \propto R^{-1}$. Therefore, there are equal contributions from all annuli binned logarithmically. The bottom line is that the vast majority of the signal is going to come from weakly lensed galaxies.

1.7 Diffuse Lenses

We've moved from a point mass to a mass distribution of a single lens, so we can now take the next step and consider distortions from all intervening mass along the line of sight. In the case of the inhomogeneous Universe, with its rich pattern of over- and under-dense regions, this is the regime of weak lensing.

Weak lensing is the same as strong lensing except that the effects are less prominent, so need to be observed statistically. For the most part, we will focus on the effect that the shapes of the background objects – mostly galaxies – are distorted. We will spend some time, though, discussing the possibility of inferring information about the distorting fields – and therefore the mass of the intervening matter – using magnification. This has become a hot topic over the past few years and is emerging as an exciting alternative to shapes. When it comes to shapes, we will show that the ellipticity of a galaxy is an excellent indicator of *cosmic shear*, which in turn is directly related to derivatives of the gravitational potentials. Schematically,

$$\epsilon_1, \epsilon_2 \to \gamma_1, \gamma_2 \to \partial^2 \Phi, \tag{1.17}$$

where the ellipticity of a galaxy is captured by two numbers, ϵ_1 and ϵ_2; the cosmic shear in the direction of the galaxy is also captured by two numbers, γ_1 and γ_2, which are related to the three possible second derivatives of the potential with respect to angular position on the sky. The magnification is a measure of the *convergence*, which is also related to derivatives of the potential. Again schematically,

$$\mu \to \kappa \to \nabla^2 \Phi, \tag{1.18}$$

where the Laplacian here is the sum of the second derivative with respect to the two angular directions on the sky, θ_1 and θ_2.

In all these cases, the signal – either a slightly more elliptical galaxy or a slightly brighter galaxy – is impossible to distinguish from the noise for a single source. The reason is that not every galaxy is circular or has the same brightness. This scatter, in shape and brightness, serves as noise, and is much larger than the signal in the case of weak lensing. Therefore, weak lensing relies on measurements of *many* background sources to beat down the noise. Beyond this, we also stop focusing on measuring the mean ellipticity, $\langle \epsilon \rangle$, since the mean is zero: different lines of sight pass through either over- or under-dense regions. Rather, we begin to focus on the variance of the fluctuations, $\langle \epsilon\epsilon \rangle$. We can get a rough estimate for the square root of the variance, or the root mean square (RMS), of these fluctuations by invoking the fact that the shear is proportional to the second derivative of the gravitational potential. To get the dimensions correct, we insert two factors of a typical cosmological distance D_L so that

$$\epsilon_{\text{RMS}} \sim D_L^2 \partial^2 \frac{\Phi_{\text{RMS}}}{c^2} \sim (D_L/\lambda)^2 \frac{\Phi_{\text{RMS}}}{c^2}, \tag{1.19}$$

where λ is the size of the structures contributing most to the deflections. Roughly $\lambda \sim 30$ Mpc and cosmological potentials are of order $10^{-5}c^2$ so we expect $\epsilon_{\text{RMS}} \sim 10^{-2}$, which turns out to be right (for the wrong reasons, as we will see later on).

Measuring cosmic shear is hard (a percent-level signal in the face of noise 30 times larger), but the potential payoff is huge: cosmic shear is a direct measurement of the matter in the Universe, regardless of whether it lights up in the form of galaxies. This is the tantalizing feature of lensing: bypassing the *tracers* of matter traditionally used to learn about structure and measuring the matter distribution directly.

As just one example of what can be done with lensing, imagine breaking up the background galaxies into coarse redshift bins, as in Fig. 1.9. The difference between the shears in the low and high redshift bins will be sensitive to the mass in more distant bins only. Doing this with multiple bins will allow us to chart the evolution of the fluctuation amplitude as a function of time.

1.8 CMB Lensing

The final step in the left–right progression in Fig. 1.3 is to generalize to sources that are not distinct objects, like galaxies. The photons that comprise the cosmic microwave background (CMB) are the prime example of this, as they arrive from all directions and hence constitute a diffuse field. The deflections each photon in the CMB experiences are small, of order arcminutes, but the structures doing the lensing are much larger, subtending degree scales. CMB lensing therefore is a bit counterintuitive: by measuring the small-scale CMB anisotropies we can infer information about the large-scale structure of the Universe.

Extracting the effect of lensing is subtle in this case of a diffuse source, and indeed the first detection of CMB lensing came after all the other effects summarized in this chapter had been detected. The secret to extracting the signal is to

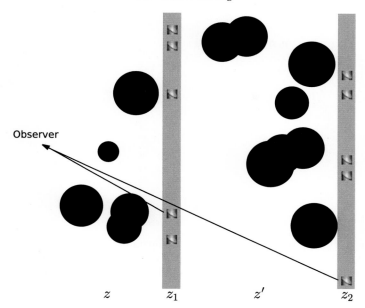

Figure 1.9 Tomography: the ellipticities of galaxies in redshift bin z_2 are sensitive to mass in the range z and z', while those in z_1 are sensitive only to matter in z. The difference in the signals therefore provides a direct measurement of the mass in z'.

gain an understanding of what the CMB would look like if it were not lensed. A combination of theory and observation has provided precisely that information. In particular, there are small differences in the intensity of the radiation in different directions: some regions of the sky have more intensity (are "hotter") and some are colder, as seen in Fig. 1.10. These variations from place to place encode information about the underlying *anisotropy spectrum*, which will occupy us in Chapter 9. The key point is that realizations of this spectrum would be the same in every region of the sky in the absence of lensing. Lensing distorts the spectrum so that it varies from place to place in the sky depending on how much mass there is along the line of sight. We will walk through the sophisticated ways scientists have developed to extract these distortions.

Roughly, though, the convergence κ can be inferred by measuring a quadratic combination of the CMB temperature field

$$T^2 \rightarrow \kappa \rightarrow \nabla^2 \Phi, \tag{1.20}$$

and from that convergence follows information about the gravitational potential. This is very similar to the information from cosmic shear; one of the main differences (and advantages) of CMB lensing is that the distance to the source is

Figure 1.10 Map of anisotropies in the temperature of the CMB by the Planck satellite. The mean temperature, $T = 2.725\text{K}$, has been subtracted off so the red and blue regions represent spots only slightly hotter and colder than average, with the typical excess temperature of these spots $\sim 80\mu\text{K}$, and the typical size degree-scale (Credit: ESA). (See color plates section.)

known exactly, thereby removing one of the big problems in cosmic shear. Indeed, another way of thinking about CMB lensing is that it adds another tomographic bin to any analysis. Like the other bins obtained from shapes or magnification, it contains information about the gravitational potential along the line of sight. Combining all this information, from a variety of observations, to extract information about physical cosmology is one of the most exciting and challenging problems in cosmology.

Suggested Reading

Some nice sources on the history of lensing are the book by Gates (2009) and the review article by Wambsganss (1998), which contains a beautiful chapter on the history. The account in §1.1 hides much of the ambiguity in the 1919 measurements. A careful look at the full story of the expeditions and data analysis was presented by Kennefick (2007).

The original article by Einstein (1936) about the dramatic effects of lensing discussed in this book begins with the sentence "Some time ago, R. W. Mandl paid me a visit and asked me to publish the results of a little calculation, which I had made at his request." The author goes on to downplay the importance of the effect, probably for two reasons. First, he had made his share of important discoveries in his life and maybe this just didn't measure up. But also Einstein, uncharacteristically, was not thinking big enough. He wrote, "Of course, there is no hope of observing this phenomenon [Einstein ring] directly," a mistake that is built on his assumption that only a star could serve as a lens. In a celebration of the 100th year of general relativity[3], the wonderful science writer Tom Siegfried filled in some of the fascinating details of the encounters between the most famous scientist in the world and Rudi Mandl, who at the time was working as a busboy washing dishes.

The astronomy basics you will need are covered in almost any undergraduate text; the one I am most familiar with is Carroll and Ostlie (2006). Extra-galactic books that come highly recommended are Schneider (2006), Serjeant (2010), and Liddle (2003).

Exercises

1.1 Compute the deflection angle given by Eq. (1.1) for a lens with the mass and radius of the Sun when the light passes by the edge of the Sun. Give your answer in arcseconds and compare it with the number in the text observed by Eddington.

1.2 Use Eq. (1.4) to determine the value of the Einstein radius of the Sun, a solar mass lens a distance 1 AU from us. To get comfortable going back and forth, express this in radians, degrees, arcminutes, and arcseconds.

1.3 The *surface brightness* is defined as the flux of energy per unit time per unit area per solid angle, so has units of Energy Time^{-1} Length^{-2} steradian^{-1}. How does the surface brightness change as an object moves farther away from us? Consider the impact of the decrease in the number of photons that reach us because we are farther away and the increase in the physical size associated with a given solid angle.

1.4 The surface density of an object is defined as

$$\Sigma(R) \equiv \int_{-\infty}^{\infty} dz\, \rho(R, z), \tag{1.21}$$

where the mass density ρ of an object at the origin is a function of the 2D radius R and the line of sight distance z in cylindrical coordinates. Calculate $\Sigma(R)$ from a tophat mass density:

$$\rho(r) = \begin{cases} \rho_0 & r < r_0 \\ 0 & r \geq r_0 \end{cases} \tag{1.22}$$

where the 3D distance $r \equiv \sqrt{R^2 + z^2}$.

---

[3] www.sciencenews.org/blog/context/amateur-who-helped-einstein-see-light

1.5 Calculate the surface density of an object with an isothermal density profile, as given in Eq. (1.15).

1.6 Assume that a given galaxy survey detects ten distant galaxies per square arcminute. Assume that they are all behind a galaxy cluster a distance 1 Gpc from us. How many background galaxies are within a projected radius of 60 kpc of the cluster center? How many within 1 Mpc from the center?

2

Deflection of Light

This chapter derives the general formulae that will be used throughout the book for the deflection of light. We start, though, with a relatively simple problem: the deflection of light by a single point mass. We will tackle this problem in two ways: first using Newtonian gravity and then using the formalism of general relativity. Both approaches find that the deflection angle is proportional to MG/bc^2 where b is the distance of closest approach, but the two theories return coefficients that differ by a factor of 2. The general relativity calculation is presented in §2.2 and §2.3. If you know general relativity, you will have no problem with those sections. If you have not yet encountered general relativity, you can skip those sections and pick up in §2.4 without missing a beat. The mathematical details in the general relativity sections, though, are designed so that even those without backgrounds in this field can work through them. The advantages of doing so are three-fold: it's always good to work through the math; you will get a glimpse of the details of many general relativity calculations and should be able to carry out others; and you might emerge with an appetite to understand the theory at a deeper level. The disadvantage is that the sections do not attempt to capture the beauty and profundity of the theory.

Armed with the result for the deflection angle, we will be able to calculate one effect exactly: the Shapiro time delay, the amount by which light is slowed down as it passes by a massive body. The rest of the chapter builds on the formula for the deflection angle and derives the lens equation, the underpinning of all gravitational lensing phenomena. Starting with this general version, where the deflection is expressed in terms of the gravitational potential, we then circle back to show how it recaptures the simple cases of a single lens and then the even simpler case of a single point mass lens.

2.1 Newtonian Deflection

Let us consider a point particle with small mass m passing by a compact object with much larger mass M. For simplicity, we will assume that the deflecting particle, the

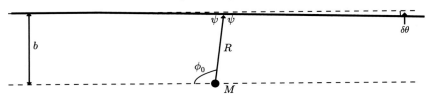

Figure 2.1 Deflection of a particle in the 2D plane as it passes by a point mass M at rest at the origin. The 2D position is parametrized by the 2D distance R and angle ϕ. The distance of closest separation in the absence of deflection, the impact parameter, is b; the deflection causes the orbit to come closest to the mass at an angle ϕ_0; and the final deflection angle is $\delta\theta$.

lens with mass M, is so heavy that its position remains fixed for the duration of the event. We choose a coordinate system with the origin at the position of the massive lens. Then, classically, Newton's equation governs the motion of the light particle, whose position is denoted $\vec{r}(t)$. If gravity is the only force acting on the particle then the relevant equation is

$$m \frac{d^2\vec{r}}{dt^2} = -\frac{mMG}{r^2}\hat{r}. \tag{2.1}$$

This is a 3D vector equation, but the motion is confined to a plane, so only two components of the equation are relevant. The distance from the deflector is denoted $R(t)$ and the angle between its current position and its position at $t = -\infty$ is ϕ, as depicted in Fig. 2.1 for the angle ϕ_0 of closest approach. We need to decompose $d^2\vec{r}/dt^2$ into $\hat{R}$ and $\hat{\phi}$ components. This is a standard exercise, and we will walk through one derivation in the next section; for now, let's simply use the result:

$$\hat{R}\left[\ddot{R} - R\dot{\phi}^2\right] + \hat{\phi}\frac{1}{R}\frac{d}{dt}\left(R^2\dot{\phi}\right) = -\frac{MG}{R^2}\hat{R}. \tag{2.2}$$

Fortuitously, the light mass m drops out of this equation, emboldening us to pretend that we have a theory that describes the motion of even a massless particle, like the photon, in the presence of gravity.

The $\hat{\phi}$ component of this equation leads to a conservation law: $R^2\dot{\phi} \equiv J_z =$ constant. Insert this conserved quantity into the radial part of the equation to get

$$\ddot{R} - \frac{J_z^2}{R^3} = -\frac{MG}{R^2}. \tag{2.3}$$

There are two tricks we can use to solve Eq. (2.3): the first is to find not the time dependence of R but rather its angular dependence; that is, find $R(\phi)$. To do this, we change the time derivatives in Eq. (2.3) to angular derivatives, which we will denote with primes. Walking slowly through the first derivative

$$\dot{R} = \frac{dR}{d\phi}\frac{d\phi}{dt} = R'\dot{\phi}$$

$$= \frac{J_z R'}{R^2} \tag{2.4}$$

and then more quickly through the second leads to

$$\ddot{R} = \frac{J_z^2}{R^2}\left[\frac{R''}{R^2} - 2\frac{R'^2}{R^3}\right]. \tag{2.5}$$

So the equation of motion for R becomes

$$\frac{R''}{R^2} - 2\frac{R'^2}{R^3} - \frac{1}{R} = \frac{-MG}{J_z^2}. \tag{2.6}$$

That was the first trick; the second is to define $u \equiv 1/R$ and rewrite the equation in terms of u. It is straightforward to show that

$$R' = \frac{-u'}{u^2}$$

$$R'' = \frac{-u''}{u^2} + \frac{2u'^2}{u^3}. \tag{2.7}$$

Plugging in leads to a remarkably simple equation for u:

$$u'' + u = \frac{MG}{J_z^2}. \tag{2.8}$$

The solution to this inhomogeneous second-order ordinary differential equation is the sum of the general solution – which we'll write as a cosine with unknown amplitude and phase – and the particular solution, which is simply given by the right-hand side. So,

$$\frac{1}{R(\phi)} = A\cos[\phi - \phi_0] + \frac{MG}{J_z^2}. \tag{2.9}$$

Before going further, let's look at the value of the conserved quantity J_z. At very early times, when the particle is still far away from the lensing mass, R is very large and ϕ is small so – to a very good approximation – equal to b/R, where b is the impact parameter. The particle moves horizontally in Fig. 2.1 so the change in ϕ is due to the change in R:

$$\dot{\phi} \simeq \frac{-b\dot{R}}{R^2}. \tag{2.10}$$

Thus initially, $J_z \equiv R^2\dot{\phi} = -b\dot{R}$. But the particle we are talking about is a photon, which travels towards the origin at the speed of light, so initially – when all the motion is horizontal – its velocity is $\dot{R} = -c$. The constant J_z therefore is equal to bc. So we can write our solution as

$$\frac{1}{R(\phi)} = A \cos\left[\phi - \phi_0\right] + \frac{MG}{b^2c^2}. \tag{2.11}$$

It remains to determine the two parameters A and ϕ_0 that are dictated by the initial conditions. The first condition is that $R \rightarrow \infty$ when $\phi = 0$, which translates into

$$0 = A \cos(\phi_0) + \frac{MG}{c^2b^2}. \tag{2.12}$$

The second condition is on the initial velocity: differentiating Eq. (2.11) with respect to time and using Eq. (2.10) leads to

$$A = \frac{1}{b \sin \phi_0}. \tag{2.13}$$

This gives us two equations to determine the two parameters A and ϕ_0. Instead of blindly cranking out the solution, let's think about the physical meaning of the angle ϕ_0. We will see that the amplitude A is positive, so the right-hand side of Eq. (2.11) is largest when $\phi = \phi_0$. Therefore, as depicted in Fig. 2.1, the particle is closest to the deflecting mass when $\phi = \phi_0$. If it were not deflected at all, the angle at which the particle would get closest to M would be $\phi = 90°$. In almost all cases of interest (e.g., see Exercises (1.1) and (2.1)), the deflection MG/bc^2 is a very small number, so ϕ_0 is only a little bit larger than $90°$. So to a good approximation, Eq. (2.13) tells us that $A = 1/b$. Plugging this into Eq. (2.12) leads to

$$\cos \phi_0 = -\frac{MG}{c^2b}. \tag{2.14}$$

This is what we expected: as depicted in Fig. 2.1, the deflection ensures that the traveling particle will get closest to the mass a bit after the undeflected $90°$ result. Writing $\phi_0 \equiv \pi/2 + \epsilon$ and Taylor expanding $\cos \phi_0 \simeq \cos(\pi/2) - \sin(\pi/2)\epsilon = -\epsilon$, we see that

$$\phi_0 = \frac{\pi}{2} + \frac{MG}{c^2b}. \tag{2.15}$$

We can now connect ϕ_0 to the deflection angle $\delta\theta$. From Fig. 2.1, we see that $2\psi + \delta\theta = \pi$ and $\phi_0 = \psi + \delta\theta$, so

$$\delta\theta_{\text{Newtonian}} = 2\phi_0 - \pi = \frac{2MG}{c^2b}, \tag{2.16}$$

the Newtonian result for the deflection angle.

2.2 Bridge

We want to compute the deflection angle in general relativity, so let's first do a warm-up exercise to get us used to the required language, in particular the metric and the geodesic equation. The exercise is to calculate the derivatives in a 2D polar coordinate system; that is, to prove the expressions we used in Eq. (2.2). The starting point is the *metric* in 2D polar coordinates, obtained by noting that a small change in distance squared is

$$ds^2 = dR^2 + R^2 d\phi^2. \tag{2.17}$$

This distance is *invariant*, meaning that two different observers choosing two different points as the origin, for example, will measure the same value for ds^2. They may measure different changes dR and $d\phi$ but these will compensate so that the physical distance measured will be the same.

The metric quantifies how small changes in the coordinates (in this case, dR and $d\phi$) lead to a small change in the distance, in this case the 2D Euclidean distance ds. Generally the relation between coordinate changes and distances is written as

$$ds^2 = \sum_{i,j=1}^{2} g_{ij} dx^i dx^j \tag{2.18}$$

where g_{ij} is the metric, in this case with three independent components g_{11}, g_{22}, and $g_{12} = g_{21}$. More generally, the metric is symmetric so a two-dimensional space requires three independent components; a 3D space, six components; and most famously, the full 4D space-time requires $16 - 6 = 10$ independent components. In this case, with $x^1 = R$ and $x^2 = \phi$, the metric is diagonal with elements that can be read off from Eq. (2.17):

$$g_{ij} = \begin{pmatrix} 1 & 0 \\ 0 & R^2 \end{pmatrix}. \tag{2.19}$$

Einstein Summation Convention

Einstein introduced a convention that is now widely used in physics. He recognized that a dot product involves summing over a single index. For example, in three dimensions, $\vec{A} \cdot \vec{B} = \sum_{i=1}^{3} A_i B_i$. The key point is that the index on both vectors is the same. In this case, then, one could recognize that the index i is to be summed over by the very fact that it appears twice. And indeed, this is very general. Typically, when an index appears twice, it is summed over. So the simple convention is to omit the summation sign and write

$$\vec{A} \cdot \vec{B} = A_i B_i \tag{2.20}$$

with the understanding that we are to sum over all values of the index i. This example is for three dimensions, but it also holds for the case of 2D polar coordinates that we are in the middle of now, and in the case of the four dimensions of space-time that we will turn to shortly. Throughout the book, we will resort to this convention and omit the summation sign when an index is repeated.

The geodesic equation for a particle can be expressed in terms of the Christoffel symbols

$$\frac{d^2 x^i}{dt^2} = -\Gamma^i{}_{jk} \frac{dx^j}{dt} \frac{dx^k}{dt} \tag{2.21}$$

where the indices i, j, k range from 1 (R) to 2 (ϕ) and from here on, we use the Einstein summation convention (see shaded box). There are two aspects of this equation that are essential ingredients of general relativity. First, there is a distinction between upper and lower indices. The metric is used to raise and lower indices. For example, $x_i = g_{ij} x^j$ and conversely, $x^i = g^{ij} x_j$, where g^{ij} with raised indices is simply the inverse of g_{ij}. In our case, with only diagonal components, the inverse is trivial and $g^{11} = 1$ and $g^{22} = 1/R^2$. The second ingredient is the Christoffel symbols. If we were working in Cartesian coordinates, the Christoffel symbols $\Gamma^i{}_{jk}$ would vanish and we would be left with the simplest form of Newton's equation: the acceleration vanishes in the absence of forces, so that $\ddot{x}^i = 0$. In polar coordinates, the Christoffel symbols encode the changing unit vectors. The simplest way to use them is to calculate them from the metric via

$$\Gamma^i{}_{jk} = \frac{g^{ii'}}{2} \left[\frac{\partial g_{i'j}}{\partial x^k} + \frac{\partial g_{i'k}}{\partial x^j} - \frac{\partial g_{jk}}{\partial x^{i'}} \right]. \tag{2.22}$$

Let's begin by calculating the geodesic equation for the R component (x^1):

$$\frac{d^2 R}{dt^2} = -\Gamma^1{}_{jk} \frac{dx^j}{dt} \frac{dx^k}{dt}. \tag{2.23}$$

But

$$\Gamma^1{}_{jk} = \frac{g^{11}}{2} \left[\frac{\partial g_{1j}}{\partial x^k} + \frac{\partial g_{1k}}{\partial x^j} - \frac{\partial g_{jk}}{\partial x^1} \right]. \tag{2.24}$$

Here, $g^{1i'}$ outside the brackets on the right has been set to g^{11} (which is equal to one) since the metric is diagonal. The g_{1i}s are all constant, equal to 1, so their derivatives vanish and we are left with

$$\Gamma^1{}_{jk} = -\frac{1}{2} \frac{\partial g_{jk}}{\partial R}. \tag{2.25}$$

Since only $g_{22} = R^2$ has an R dependence, the only relevant nonzero Christoffel symbol is

$$\Gamma^1{}_{22} = -R. \tag{2.26}$$

So the geodesic equation for R – in the absence of gravity – is

$$\ddot{R} - R\dot{\phi}^2 = 0 \tag{2.27}$$

in agreement with Eq. (2.2).

2.3 Deflection from a Point Mass in General Relativity

We can use the formalism developed above to compute the deflection of light by a point mass in general relativity. In the case of polar coordinates, space itself was Euclidean but the coordinates we were using to describe the space were complicated, and the geodesic equation was able to deal with those complications. We now treat a space that is distorted by gravity so it is *curved*. The familiar example is the surface of the Earth, where the same change in longitude at the equator corresponds to a larger change in distance than an identical longitude change near the poles. The metric captures this subtlety and the geodesic equation is still the ideal way to track how these complications affect the motion of a particle.

We are interested in understanding how the path of a (massless) particle, tracked by coordinates $x^i \equiv (x(t), y(t), z(t))$, is distorted by a gravitational field. We will choose the axes so that the photon begins its motion traveling along the z-axis. *Distortion* then means that it will acquire a small nonzero value of $x^1 = x$ or $x^2 = y$. We want to apply the geodesic equation then to one of these transverse directions:

$$\frac{d^2 x^i}{d\lambda^2} = -\Gamma^i{}_{\alpha\beta} \frac{dx^\alpha}{d\lambda} \frac{dx^\beta}{d\lambda}. \tag{2.28}$$

Note that there are several differences from the simple case we examined above. We are now dealing with all three spatial and one temporal dimension, so the Greek indices α, β run from 0 (for time) to 3. Latin indices i, j will be used for the two dimensions transverse to the line of sight ($x^1 = x$; $x^2 = y$). Another difference is that this relativistic version of the geodesic equation cannot give time t a special role, so the derivatives are not with respect to time but rather with respect to a monotonically increasing parametrization of the path, a so-called *affine parameter* λ. The simplest way to relate λ to things we understand is to normalize it so that the derivative of the four-vector x^α with respect to λ is equal to the relativistic four-momentum vector:

$$\frac{dx^\alpha}{d\lambda} \equiv p^\alpha = (E/c, \vec{p}) \tag{2.29}$$

where c is the speed of light. So the derivatives on the right-hand side of Eq. (2.28) can be simply replaced by p^α and p^β. Further, the derivatives with respect to λ on the left-hand side of Eq. (2.28) can be transformed into derivatives with respect to time. The first derivative becomes

$$\frac{dx^i}{d\lambda} = \frac{dx^i}{dt} \frac{dt}{d\lambda} = \frac{E}{c} \frac{dx^i}{dt}. \tag{2.30}$$

Then the second derivative is

$$\frac{d^2x^i}{d\lambda^2} = \frac{E}{c}\frac{d}{dt}\left[\frac{E}{c}\frac{dx^i}{dt}\right] \simeq \left(\frac{E}{c}\right)^2 \frac{d^2x^i}{dt^2}. \tag{2.31}$$

The approximate equality here encodes some physics: the energy does change with time since the particle moves in and out of gravitational potentials, but the change is small. Since the transverse deflection x^i is also first-order in this small perturbation to the path, including any change in E would be keeping a second-order effect. We will work to first order only, so pass E through the derivative. We are thus left with

$$\frac{E^2}{c^2}\frac{d^2x^i}{dt^2} = -\Gamma^i{}_{\alpha\beta}p^\alpha p^\beta. \tag{2.32}$$

It remains to compute the Christoffel symbols, which in turn requires us to specify the metric. Let's focus on the deflection from a point source with mass M so that we are redoing the classical problem from §2.1. The gravitational effect of the point mass translates into a distorted metric:

$$g_{00} = c^2\left(1 - \frac{2MG}{rc^2}\right)$$

$$g_{ij} = -\delta_{ij}\left(1 + \frac{2MG}{rc^2}\right). \tag{2.33}$$

Here, the Kronecker delta δ_{ij} is defined to be equal to one if $i = j$ and zero otherwise. If MG were zero, this would be the *Minkowski metric* (Exercise (2.4)). With nonzero MG, the first of these equations captures the effect of *time dilation*, that clocks move more slowly in a gravitational well. To see this, consider two events at the same location ($dx = 0$) so that the invariant distance is $ds = dt\sqrt{g_{00}}$. Therefore, the coordinate time elapsed as measured by an observer will be equal to

$$dt = \frac{ds}{c\sqrt{1 - \frac{2MG}{rc^2}}} \simeq \frac{ds}{c}\left[1 + \frac{MG}{rc^2}\right]. \tag{2.34}$$

So the elapsed time will be larger near a massive object, the defining feature of time dilation. The second line in Eq. (2.33) says that an observer in a deep potential well will measure distances to be much smaller than will an observer far from the mass. Quantitatively, two events at the same time will be separated by a coordinate difference dx approximately equal to $ds \times (1 - MG/rc^2)$: this captures the phenomenon of *length contraction*.

We can use this metric in Eq. (2.22) to calculate the Christoffel symbols; the component with an upper spatial index and two lower time indices is

$$\Gamma^i{}_{00} = \frac{1}{2} g^{ii'} \left[\frac{\partial g_{i'0}}{\partial x^0} + \frac{\partial g_{i'0}}{\partial x^0} - \frac{\partial g_{00}}{\partial x^{i'}} \right] \tag{2.35}$$

where Latin indices i, i' are spatial, so run from 1–3. The first two terms in brackets vanish because there are no off-diagonal elements of the metric. Carrying out the derivative in the last term and retaining only terms linear in MG/rc^2 leads to

$$\Gamma^i{}_{00} = \frac{MGx^i}{r^3}. \tag{2.36}$$

We can now evaluate the right-hand side of the geodesic equation:

$$-\Gamma^i{}_{\alpha\beta} p^\alpha p^\beta = -\frac{MGx^i}{r^3} \frac{E^2}{c^2} - 2\Gamma^i{}_{0j} p^0 p^j - \Gamma^i{}_{jk} p^j p^k. \tag{2.37}$$

The other symbols are easily computed (see Exercise (2.5)). The middle term can be neglected because it is second order in the small deflection parameter: $\Gamma^i{}_{0j}$ is first order in MG/rc^2, as is the transverse momentum p^j. The final term contributes only when $j = k = 3$; then it is equal to $-(p^z)^2 x^i MG/r^3 \simeq -E^2 x^i MG/r^3 c^2$, so it is equal to the first term (again p^z is equal to E/c at zeroth order). We are then left with a simple equation for the deflection:

$$\frac{d^2 x^i}{dt^2} = -\frac{2MGx^i}{r^3}. \tag{2.38}$$

Eq. (2.38) is identical to Eq. (2.1), the Newtonian equation that governs motion, apart from the factor of 2. So turning the crank as we did in the steps leading to Eq. (2.16) leads immediately to the prediction that

$$\delta\theta_{GR} = \frac{4MG}{c^2 b}. \tag{2.39}$$

The deflection angle predicted by general relativity is twice as large as the Newtonian prediction.

2.4 Shapiro Time Delay

We now have enough ammunition for a brief diversion to compute the Shapiro time delay: the amount by which light is delayed as it lingers in the potential well around a massive object. For the massless photons that comprise light rays, the invariant distance squared

$$ds^2 = g_{00}dt^2 + g_{ij}dx^i dx^j = 0. \tag{2.40}$$

Imagine light rays traveling along the z-axis (so that dx^i is nonzero only for $i = 3$) passing near the Sun, so that the metric is given by Eq. (2.33) with $M = M_\odot$. In a small time dt they traverse a small coordinate distance dz given by

$$dz = c\,dt\left[\frac{1 - 2MG_{\odot}/rc^2}{1 + 2MG_{\odot}/rc^2}\right]^{1/2}$$

$$\simeq c\,dt\left[1 - \frac{2MG_{\odot}}{rc^2}\right]. \tag{2.41}$$

In the absence of the gravity, the speed of light dz/dt would simply be equal to c, but gravity slows the light down, so that $dz/dt < c$. Irwin Shapiro, in 1964, proposed an ingenious way of observing this effect. If we shoot a laser pulse past the limb of the Sun at the time when a planet, say Venus, is directly behind the Sun, some of the light will rebound back to us, and we could measure the total travel time.

To obtain the total travel time T from the Earth to Venus, we need to divide both sides of Eq. (2.41) by $1 - 2MG_{\odot}/rc^2$, Taylor expand again, and then integrate:

$$\int_{-D_E}^{D_V} dz\left[1 + \frac{2MG_{\odot}}{rc^2}\right] = c\int_0^T dt. \tag{2.42}$$

The coordinate system is set up as in Fig. 2.2, with the Sun at $z = 0$, the light ray starting at Earth a distance D_E away and hitting a planet (in this case Venus) and then bouncing back. The integral is along the z-direction so the factor of r in the integrand needs to be expressed as $r = \sqrt{z^2 + R_{\odot}^2}$. Therefore, the amount by which the light ray is delayed is

$$\Delta t_{\mathrm{Sh}} = \frac{2}{c^3}\int_{-D_E}^{D_V} dz\,\frac{2MG_{\odot}}{\sqrt{z^2 + R_{\odot}^2}} \tag{2.43}$$

where the factor of 2 in front accounts for both segments of the round trip.

Carrying out the integral leads to

$$\Delta t_{\mathrm{Sh}} = \frac{4MG_{\odot}}{c^3}\ln\left[z + \sqrt{z^2 + R_{\odot}^2}\right]_{-D_E}^{D_V}$$

$$\simeq \frac{4MG_{\odot}}{c^3}\ln\left[\frac{4D_V D_E}{R_{\odot}^2}\right] \tag{2.44}$$

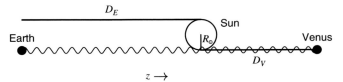

Figure 2.2 Light ray emitted from Earth, passing by the Sun, bouncing off Venus, and returning to Earth. The solar radius is denoted $R_{\odot}$ and the distances between the Sun and the two planets are D_E and D_V respectively. Not drawn to scale.

where the last line follows from the fact that the solar radius is much smaller than the planet–Sun distances. Plugging in numbers ($D_V = 1.1 \times 10^8$ km; $D_E = 1.5 \times 10^8$ km; $R_\odot = 4.3 \times 10^5$ km) leads to

$$\Delta t_{Sh} = 200\mu s \tag{2.45}$$

for the case when Venus is at opposition.

The time delay predicted by Shapiro (1964) was first detected by sending pulses of light to Venus when its line of sight passed close to the Sun (Shapiro et al., 1968). Since then a number of targets, natural and man-made, have been used, and the measurements have become ever more precise. One way to quantify this is to place an arbitrary amplitude in front of the right-hand side of Eq. (2.44), traditionally called $(1 + \gamma)/2$, whose value is equal to 1 in general relativity. Fig. 2.3 shows a

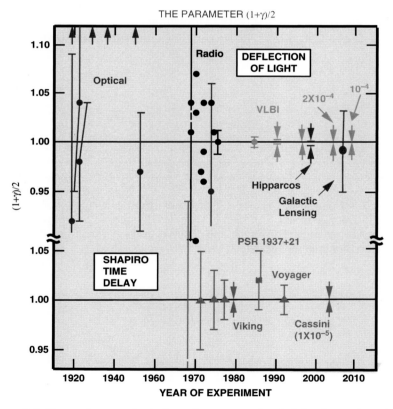

Figure 2.3 Constraints on deviations from general relativity from time delay experiments from Will (2014). The bottom half of the figure depicts constraints from detections of the time delay signal, such that the general relativity prediction corresponds to amplitude $(1 + \gamma)/2 = 1$, as discussed in the text. The top half shows constraints from the deflection of light, such as the first detection during the Solar eclipse of 1919. (See color plates section.)

number of experiments since the first detection and the increasingly precise limits
on this "post-Newtonian" parameter. While the initial measurement was accurate
at the 10 percent level, the Cassini spacecraft constrained $(1 + \gamma)/2$ to be equal to
the value predicted in general relativity to one part in one hundred thousand.

2.5 Angular Vectors

Until now, we have been thinking about deflection as it would be seen if we were
riding on the light particles. Let's shift to the more realistic vantage point where we
are the observers, say at the origin, viewing light that has traveled from a distant
source. The distance to the source is D_S and Fig. 2.4 shows the geometry.

Although not obvious from the figure, the angles have both an amplitude and a
direction. The amplitude describes how the incoming ray is tilted with respect to the
z-axis, where the line connecting the observer to the lens is chosen to be the z-axis.
The directions of the angles specify the locations in the x–y plane (perpendicular
to the page in Fig. 2.4). Fig. 2.5 shows these angles with amplitudes and directions

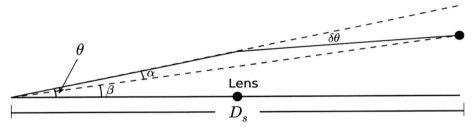

Figure 2.4 Light from a source a distance D_S from us would be observed at an
angle β if not deflected. The light is deflected, though, so the source appears to
us to reside an angle θ from the line connecting the observer and the lens. The
difference between the two angles is α.

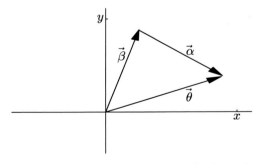

Figure 2.5 Angular vectors in the x–y plane perpendicular to the line of sight (the
plane of the sky). The true source position is $\vec{\beta}$, while the observed image is at $\vec{\theta}$.
The deflection angle α relates the two.

in the 2D plane. So the transverse position of the ray propagated backwards relative to the z-axis is $D_S \vec{\theta}$ and the point therefore has 3D coordinates $(D_S \vec{\theta}, D_S)$. Since the light has been deflected, this is *not* the true source position; let's denote the true transverse position in the $z = D_S$ plane as $D_S \vec{\beta}$. Then, $\vec{\beta}$ is the true (unlensed) angular position of the source, while $\vec{\theta}$ is the position we ascribe to it since we see the light coming to us from that direction. As depicted in Fig. 2.5, the vector difference between them – the deflection angle – is called $\vec{\alpha}$. We will see that there is a small difference between $\vec{\alpha}$ and what we have been calling the deflection angle until now, $\delta\vec{\theta}$. From now on, we reserve the term for $\vec{\alpha}$.

2.6 Deflection by the Gravitational Potential

The point mass deflector treated in §2.3 is often a very bad approximation. As light travels through space, it experiences a variety of deflections due to the fluctuating gravitational potential along the line of sight. We now generalize the previous discussion to write down the formula that will account for lensing in a wide variety of circumstances.

The starting point will be to tweak Eq. (2.38) in two ways. First, let's notice that the right-hand side can be written in terms of the gravitational potential as

$$-\frac{2MGx^i}{r^3} = 2\frac{\partial(MG/r)}{\partial x^i} \rightarrow -2\frac{\partial\phi}{\partial x^i}. \qquad (2.46)$$

Indeed the metric given by Eq. (2.33) describes a space-time perturbed by a point mass, but the more general version simply replaces MG/r with $-\phi$, so if we walked through the derivation again we would obtain a right-hand side equal to twice the derivative of the potential. The second tweak will be to replace the time derivatives on the left of Eq. (2.38) with derivatives with respect to z, the main direction of propagation. This is not strictly correct, as there are changes with time besides those along the line of sight, but these changes are small, so if we kept them we would be keeping a second-order term. Since we are working to first order in the deflection only, moving to derivatives with respect to z/c is fine. We are then left with

$$\frac{d^2 x^i(z)}{dz^2} = -\frac{2}{c^2}\frac{\partial\phi(x^i, z; t)}{\partial x^i}. \qquad (2.47)$$

Recall that this is an equation for the transverse components of the position, so i ranges from 1–2, with $x^3 = z$. Note that the sign here is correct: an overdensity at the origin means the potential is at a minimum at the origin, so its derivative is positive when x^i is slightly greater than zero. The positive derivative means that the force is negative, inwards towards the over-dense region. We want to write this as an equation for the angle $\vec{\theta}$ between the photon and the z-axis. At every value of z, the transverse distance $x^i = z\theta^i$, so

$$\frac{d^2(z\theta^i(z))}{dz^2} = -\frac{2}{c^2}\frac{\partial\phi(x^i, z; t)}{\partial x^i}. \tag{2.48}$$

Let's start the light ray out at our position (chosen to be the origin with $x = y = z = 0$) leaving with an angle $\vec{\theta}$ and trace it back to its source in $z = D_S$ plane. Integrate the lens equation once from $z = 0$ out to z' so that

$$\frac{d(z'\theta^i(z'))}{dz'} = C^i - \frac{2}{c^2}\int_0^{z'} dz \frac{\partial\phi(x^i = z\theta^i, z; t)}{\partial x^i}. \tag{2.49}$$

We will get to the constant C^i shortly. Integrate again this time out to $z' = D_S$ and use the fact that the true angular position at $z' = D_S$ is $\vec{\beta}$ to get

$$D_S\beta^i = D_SC^i - \frac{2}{c^2}\int_0^{D_S} dz' \int_0^{z'} dz \frac{\partial\phi(x^i = z\theta^i, z; t)}{\partial x^i}. \tag{2.50}$$

The time in the integrand is simply set by the position: $t = t_0 - z/c$ where t_0 is the time today.

Consider the 2D integral in Eq. (2.50). Although it is over the shaded region in Fig. 2.6, the integrand depends only on z, so clearly it is simplest to first integrate trivially over z' and then over z. To switch the order of integration, note from the figure that z' ranges from z to D_S. This becomes the range of the inner integral, while the limits on the now outer z integral are from 0 to D_S. Therefore,

$$\beta^i = C^i - \frac{2}{c^2 D_S}\int_0^{D_S} dz \frac{\partial\phi(z\theta^i, z; t)}{\partial x^i}\int_z^{D_S} dz'$$

$$= C^i - \frac{2}{c^2 D_S}\int_0^{D_S} dz \frac{\partial\phi(z\theta^i, z; t)}{\partial x^i}(D_S - z). \tag{2.51}$$

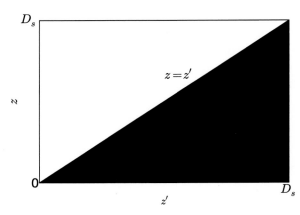

Figure 2.6 The region of integration for the 2D integral in Eq. (2.50) that determines the lensing deflection.

To determine the integration constant C^i, consider its value when there is no deflection: in that case $\vec{\beta}$ is just equal to the apparent position $\vec{\theta}$, so that is the value of the constant. We are thus left with the *lens equation*:

$$\vec{\beta} = \vec{\theta} - \vec{\alpha}(\vec{\theta}) \tag{2.52}$$

where the 2D deflection angle $\vec{\alpha}$ has components

$$\alpha^i(\vec{\theta}) \equiv \frac{2}{c^2 D_S} \int_0^{D_S} dz \, (D_S - z) \, \frac{\partial \phi(z\vec{\theta}, z; t = t_0 - z/c)}{\partial x^i}. \tag{2.53}$$

We will see that the lens equation, while compact, hides quite a bit of complexity. Think of the right-hand side of Eq. (2.52) as a function of $\vec{\theta}$ in the 2D plane. There are potentially many different values of $\vec{\theta}$ for which the right-hand side is equal to $\vec{\beta}$, so there are potentially many different images of a single source.

The deflection vector depends on the position on the sky $\vec{\theta}$ and is an integral along the line of sight defined by this position back to the source. Another way of writing this stems from noting that the derivative with respect to the transverse coordinates x^i can be written as $(1/z)\partial/\partial\theta^i$ since the transverse coordinates are equal to $z\theta^i$. Then the derivative with respect to θ^i can be brought outside the integral so that the deflection angle is equal to

$$\alpha^i(\vec{\theta}) = \frac{\partial \Phi(\vec{\theta})/c^2}{\partial \theta^i}, \tag{2.54}$$

where the *projected gravitational potential* Φ is defined as

$$\Phi(\vec{\theta}) \equiv \frac{2}{D_S} \int_0^{D_S} dD_L \, \phi \left(x^i = D_L \theta^i, D_L; t = t_0 - D_L/c \right) \frac{D_{SL}}{D_L}. \tag{2.55}$$

Here the dummy variable z has been replaced with D_L, the distance between us and any mass along the line of sight that serves as a lens, and the distance between the source and the lens is defined as D_{SL}. The last argument of the 3D potential indicates that it is to be evaluated at the time t at which the light passed by the plane at D_L (t_0 is the time today). Eq. (2.55) gives an indication of which structures will contribute most to the deflection. In particular, the contribution to Φ from inhomogeneities close to the source is suppressed by the D_{SL} factor. More generally, if the potential is constant on the sky, then there is no deflection. Lensing then emerges from changes in the projected gravitational potential across the sky.

There is one other feature hidden in Eq. (2.55): an approximation that was implicit in our derivation. The integral along the line of sight samples the potential at the transverse position $D_L\vec{\theta}$. This is *not* the actual path taken by the light ray as illustrated in Fig. 2.7. So the decision to evaluate the 3D potential along the dashed curve in Fig. 2.7 is an approximation, known as the Born approximation. If we tried to be more careful and Taylor expanded $\phi(\vec{x}^{\text{true}})$ around the dashed

Image

Source

Figure 2.7 The true path taken by the light ray from the source (jagged curve) compared to the line of sight back to the image inferred from the arrival direction. Evaluating the potential along the dashed line is called the Born approximation.

path $D_L \vec{\theta}$, we would obtain a term proportional to $\partial \phi$ multiplied by the difference between the true path and the apparent one. This difference, between the dashed curve (apparent path) and jagged curve (real path), is itself proportional to variations in the gravitational potential. So the correction term would be second-order in the potential and so very small. The Born approximation is sufficient for almost all applications.

2.7 Thin Lens Approximation

Now that we have set up the formalism to attack deflection by a general gravitational field, let's return to the special case of a single lens, where most of the deflection is due to a lens at one fixed region along the line of sight. This almost, but not quite, brings us back to the point mass case of §2.3. The similarity is that both sections treat single lenses; the difference is that the lens in this section can be extended in the transverse direction. This extension turns out to be extremely important in a wide number of cases. Also, the special case treated here gives us a chance to get comfortable using the formalism developed in the previous section.

As a first step, we invoke the solution to Poisson's equation, which relates the three-dimensional gravitational potential to the mass density:

$$\phi(\vec{x}) = -G \int \frac{d^3 x'}{|\vec{x} - \vec{x}'|} \rho(\vec{x}'). \tag{2.56}$$

This, together with Eq. (2.55), means that the projected potential, whose derivative determines the deflection angle, can be written as an integral over four dummy variables: $\vec{x}'$ here and the line of sight distance in Eq. (2.55). To obtain $\Phi(\vec{\theta})$, we need to evaluate the 3D potential ϕ at $\vec{x} = (D'_L \vec{\theta}, D'_L)$, where we use D'_L as the dummy variable in Eq. (2.55) to distinguish it from the single distance D_L that will dominate the integral. Breaking up the $d^3 x'$ integral into a radial and transverse part leads then to

$$\Phi(\vec{\theta}) = \frac{-2G}{D_S} \int_0^{D_S} dD'_L \frac{D'_{SL}}{D'_L} \int \frac{d^2 x_\perp \, dz}{\sqrt{(\vec{\theta} D'_L - \vec{x}_\perp)^2 + (D'_L - z)^2}} \rho(\vec{x}_\perp, z). \tag{2.57}$$

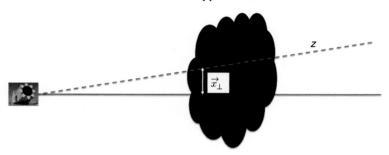

Figure 2.8 The surface density is an integral of the mass density along the line of sight z. It is a function of the perpendicular distance from the z-axis connecting the lens to the observer.

The assumption of a single lens means that the density is sharply peaked at some value $z = D_L$. Therefore, in the denominator in Eq. (2.57), where the dependence on z is mild, it is an excellent approximation to replace z with the value D_L. Then the z integral simplifies to include only the density $\rho(\vec{x}'_\perp, z)$, and we define that integral as the *surface density* as depicted in Fig. 2.8:

$$\Sigma(\vec{x}_\perp) \equiv \int dz\, \rho(\vec{x}_\perp, z). \tag{2.58}$$

The surface density then varies across the field and represents the density integrated along the line of sight at a particular value of $\vec{x}_\perp$. The projected potential is then

$$\Phi(\vec{\theta}) = \frac{-2G}{D_S} \int_0^{D_S} dD'_L \frac{D'_{SL}}{D'_L} \int d^2 x_\perp \frac{\Sigma(\vec{x}_\perp)}{\sqrt{(\vec{\theta} D'_L - \vec{x}_\perp)^2 + (D'_L - D_L)^2}}. \tag{2.59}$$

To make progress, we change orders of integration and work on the D'_L integral first. The same physical principle holds: by assumption, the largest contribution to the integral will be when $D'_L = D_L$, so we replace all relatively slowly varying functions of D'_L with their values at $D'_L = D_L$. Here, this impacts the D'_{SL}/D'_L factor and the $D'_L\vec{\theta} - \vec{x}'_\perp$ in the denominator, so that

$$\Phi(\vec{\theta}) \simeq \frac{-2GD_{SL}}{D_S D_L} \int d^2 x_\perp\, \Sigma(\vec{x}_\perp) \int_0^{D_S} \frac{dD'_L}{\sqrt{(\vec{\theta} D_L - \vec{x}_\perp)^2 + (D'_L - D_L)^2}}. \tag{2.60}$$

The inner integral is $\ln[D'_L - D_L + \sqrt{(D'_L - D_L)^2 + (\vec{\theta} D_L - \vec{x}_\perp)^2}]$. When this is evaluated at the upper ($D'_L = D_S$) and lower limits, the transverse distance $(\vec{\theta} D_L - \vec{x}_\perp)$ will be much smaller than the other term in the square root, either D_L or D_{SL}. Therefore, the upper limit term becomes simply $\ln(2D_{SL})$ while the leading terms cancel in the lower limit, so a Taylor expansion leads to the conclusion that the argument of the log is $[\vec{\theta} D_L - \vec{x}_\perp]^2/2D_L$. The projected potential is then

$$\Phi(\vec{\theta}) \simeq \frac{-2GD_{SL}}{D_S D_L} \int d^2 x_\perp \, \Sigma(\vec{x}_\perp) \, \ln\left[\frac{4D_{SL}D_L}{(\vec{\theta} D_L - \vec{x}_\perp)^2}\right]. \tag{2.61}$$

We, though, are interested only in the part of this that depends on $\vec{\theta}$ since we will eventually differentiate with respect to $\vec{\theta}$, so the numerator of the log can be ignored and the projected potential becomes

$$\Phi(\vec{\theta}) \rightarrow \frac{4GD_{SL}}{D_S D_L} \int d^2 x'_\perp \, \Sigma(\vec{x}'_\perp) \, \ln\left|\vec{\theta} D_L - \vec{x}'_\perp\right|. \tag{2.62}$$

It is a bit cleaner to write the integral as one over the observed *angular* distribution of the surface density. So define a new angular dummy variable $\vec{\theta}' \equiv \vec{x}'_\perp/D_L$. Then the potential is

$$\Phi(\vec{\theta}) = \frac{4GD_{SL}D_L}{D_S} \int d^2\theta' \, \Sigma(\vec{\theta}') \, \ln\left|\vec{\theta} - \vec{\theta}'\right|. \tag{2.63}$$

The prefactor dictates that a lens will have maximum impact when it is not too close to the observer (i.e., the signal grows as D_L gets larger) and when it is not close to the source (it goes to zero as D_{SL}). So a lens halfway in between the source and the observer produces the most dramatic effects. We also see that the density contributes to the potential at a given position weighted by the log of the distances between the density element and the position.

The approximation that led to Eq. (2.63) is known as the *thin lens approximation*. The thin lens expression leads to two important definitions. Consider the prefactor in Eq. (2.63): when divided by c^2, as is done to obtain the deflection angle, it has dimensions of (Length^2/Mass) or the inverse of the dimensions of the surface density. It makes sense, therefore, to define the critical surface density as

$$\Sigma_{\text{cr}} \equiv \frac{c^2 D_S}{4\pi G D_{SL} D_L}. \tag{2.64}$$

Then the surface density can be expressed in units of the critical surface density

$$\kappa(\vec{\theta}) \equiv \frac{\Sigma(\vec{\theta})}{\Sigma_{\text{cr}}}. \tag{2.65}$$

This dimensionless κ is also called the *convergence*, and it will maintain importance even in the non-thin lens case. With these definitions, the deflection angle for a single thin lens becomes simply

$$\alpha^i \Big|_{\text{Thin Lens}} = \frac{\partial}{\partial \theta^i} \left[\frac{1}{\pi} \int d^2 \theta' \, \kappa(\vec{\theta}') \, \ln \left| \vec{\theta} - \vec{\theta}' \right| \right]$$

$$= \frac{1}{\pi} \int d^2 \theta' \, \kappa(\vec{\theta}') \, \frac{\theta^i - \theta'^i}{|\vec{\theta} - \vec{\theta}'|^2}. \tag{2.66}$$

Several points about Eq. (2.66): first, the angular factor in the integrand explains how the contributions to the deflection at position $\vec{\theta}$ are weighted. Density far from the angle produces a deflection along the line connecting the density and the position inversely weighted by the angular distance between them. Second, everything here is dimensionless since transverse distances have been expressed in terms of angles and the density in units of the critical density. This offers a clue that lensing will be a *strong* effect if κ is greater than unity and will be *weak* if κ is less than unity. And indeed, this nomenclature has been adopted to delineate the regions of strong lensing and weak lensing.

2.8 Back to the Point Mass

Let's complete the circle by using our new general formula to compute the deflection from a point mass. A point mass has surface density equal to $\Sigma = M \delta_D^2(\vec{\theta})/D_L^2$ where $\delta_D^2()$ is the two-dimensional Dirac delta function, which will simplify many of our calculations (Exercise (2.7) is useful if you have never worked with these). For the integral over the surface density in Eq. (2.63), the Dirac delta function sets $\vec{\theta}'$ equal to zero everywhere in the integrand, so the projected potential is

$$\Phi(\theta) = \frac{4MGD_{SL}}{D_S D_L} \ln |\theta| \tag{2.67}$$

and the deflection angle becomes

$$\vec{\alpha}(\vec{\theta}) \Big|_{\text{point mass}} = \frac{4MGD_{SL}}{c^2 D_S D_L} \frac{\vec{\theta}}{\theta^2}. \tag{2.68}$$

This is not obviously equal to the answer we got in §2.3. The direction part of it is extra, but it agrees with what we implicitly assumed before: if the incoming ray lay on the x-axis in the 2D plane perpendicular to propagation, then the deflection would be along that axis. This is consistent with the initial result. The amplitude is *not* equal to $4MG/D_L\theta c^2$, which was our previous result. The reason for this small discrepancy can be gleaned by looking again at Fig. 2.4. We have computed α here, but what was computed above was $\delta\theta$. You will show in Problem 2.8 that the two angles are related in such a way that our expression for α here agrees with our earlier expression for $\delta\theta$.

We will take up multiple images in the next chapter, but it is too tempting to pass up the opportunity to glimpse at the lens equation now in the case of a point mass. It becomes

$$\vec{\beta} = \vec{\theta} - \frac{4MGD_{SL}}{c^2 D_S D_L} \frac{\vec{\theta}}{\theta^2}. \tag{2.69}$$

Again, leaving the details for the next chapter, we can still note that when the source is directly behind the lens ($\vec{\beta} = 0$), the right-hand side must vanish. This happens for all θ satisfying

$$\theta = \theta_E \equiv \sqrt{\frac{4MGD_{SL}}{c^2 D_S D_L}}. \tag{2.70}$$

Therefore, we observe a circular ring centered at the actual position of the source with radius equal to θ_E.

Suggested Reading

The short introduction to general relativity here is meant only for those who want to use it to get to lensing results. To delve deeper, there are several wonderful books at this level: "A First Course in General Relativity" (Schutz, 2009) is great, and I've heard equally strong praise for the more recent book by Hartle (2003). On a more advanced level, Sean Carroll's book (Carroll, 2004) is excellent.

There are several excellent textbooks on gravitational lensing, most pitched at a slightly more formal level than this one. Schneider et al. (1992) was written over twenty years ago but still holds up because it contains a rigorous treatment of the fundamental aspects of lensing. A more recent set of lectures is in Schneider et al. (2006). A short, but very informative, book was written by Mollerach and Roulet (2002); it too covers the basics and also many illuminating figures, especially in the area of strong lensing.

Shapiro time delays were first introduced in Shapiro (1964) and followed up soon afterwards with a detection in Shapiro et al. (1968). Will (2014) is a nice recent review. The standard book about experimental tests of general relativity is "Theory and Experiment in Gravitational Physics" (Will, 1993).

Exercises

2.1 Compute the deflection angle given by Eq. (2.14) for a galaxy cluster lens with mass equal to $10^{14} M_\odot$ when the light passes 100 kpc from the cluster center.

2.2 Typical deflection angles of interest in lensing are of order the Einstein radius $\theta_E \equiv [4MG/Dc^2]^{1/2}$. Show that the Einstein radius is much larger than the angle subtended by the Schwarzschild radius, $R_s \equiv 2MG/c^2$. Exercises (1.1) and (2.1) indicate that, in most cases of interest, the light detected in lensing effects passes far from the Schwarzschild radius.

2.3 Starting from the $i = 2$ component in Eq. (2.21), show that the geodesic equation for ϕ agrees with that in Eq. (2.2).

2.4 The Minkowski metric has $g_{00} = c^2$ and $g_{ij} = \delta_{ij}$ Show that $p^\alpha p_\alpha \equiv g_{\alpha\beta} p^\alpha p^\beta = 0$ for massless particles in the Minkowski metric.

2.5 Show that

$$\Gamma^i{}_{0j} = 0$$

$$\Gamma^i{}_{jk} = -\frac{MG}{r^3 c^2} \left[\delta_{ij} x^k + \delta_{ik} x^j - \delta_{jk} x^i \right] \tag{2.71}$$

for the metric in Eq. (2.33).

2.6 From Eq. (2.63) in the thin lens approximation, show that $\nabla^2 \Phi = 2c^2\kappa$, where $\nabla^2 \equiv \partial^2/\partial\theta_x^2 + \partial^2/\partial\theta_y^2$. This suggests a more general definition for the convergence: κ can be defined as a solution to the 2D Poisson equation even when the thin lens approximation breaks down.

2.7 Consider the following definition of the one-dimensional Dirac delta function

$$\int_{-\infty}^{\infty} dx \, f(x) \, \delta_D(x) = f(0). \tag{2.72}$$

By changing dummy variables from $x \to \lambda x$, show that

$$\int_{-\infty}^{\infty} dx \, f(x) \, \delta_D(\lambda x) = \frac{f(0)}{\lambda}. \tag{2.73}$$

More generally, show that, when dealing with two-dimensional Dirac delta functions,

$$\int_{-\infty}^{\infty} d^2x \, f(\vec{x}) \, \delta_D(\lambda\vec{x} - \lambda\vec{x}_0) = \frac{f(\vec{x}_0)}{\lambda^2}. \tag{2.74}$$

2.8 Using Fig. 2.4, show that

$$\alpha = \frac{D_{SL}}{D_S} \delta\theta. \tag{2.75}$$

3

Multiple Images

Deflection of light allows for the fascinating possibility that two rays emanating from the same source can travel along different paths and reach the same destination. We observers would then see two images of the same object arriving from two different directions: that is, we would see two copies of the same object in different places on the sky. And it is not just two: in principle, there could be many different images of the same object. The fundamental equation governing the number of images is the lens equation, Eq. (2.52):

$$\vec{\beta} = \vec{\theta} - \vec{\alpha}(\vec{\theta}) \tag{3.1}$$

where $\vec{\beta}$ is the true source position; $\vec{\theta}$ the observed position; and $\vec{\alpha}$ the deflection angle, all as depicted in Fig. 2.5. The mass distribution of the lens(es) determines the deflection angle $\vec{\alpha}$. The observed images then are not only spectacular but also provide information about the mass distribution of the foreground matter.

One way of thinking about the lens equation, one that will become useful starting with the examples in this chapter, is as a mapping from $\vec{\beta}$, the true position of the source to the images $\vec{\theta}$. This is said to be a mapping from the *source plane* to the *image plane*. The statement that a lens can produce multiple images is equivalent to the statement that this mapping from the source plane to the image plane is not necessarily one-to-one.

On a less mathematical level, consider just how astonishing the phenomenon of multiple images can be. Fig. 3.1 shows four images of the same supernova, named SN Refsdal. Supernovae are among the brightest objects in the universe, and this one happens to lie near the line of sight connecting us to an intervening galaxy. The galaxy served as a lens and produced four images of the same event. Perhaps the most remarkable thing about this lensed supernova is that each image reached us at a slightly different time because the light in each took a different path. Although the time differences between the four images shown in Fig. 3.1 were too small to

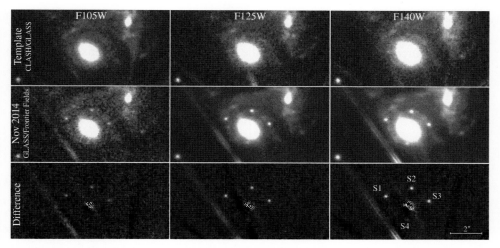

Figure 3.1 Images of a lensed supernova (Kelly et al., 2015). The columns show the images taken by the Hubble Space Telescope in three different wavelength bands. The top row is the image before the supernova exploded; the middle row captures the field when the images of the supernova become prominent; and the bottom row shows the difference between the before and during images. Note the existence of four images of the same supernova.

be detected, modelers have shown that another image *of the same supernova* is expected to appear on the timescale of about a year!

3.1 Point Mass Lens

The simplest possibility is that the lens is a point mass, so that the only free parameter is the mass of the lens. In that case, the deflection angle is given by Eqs. (2.68) and (2.70) to be

$$\vec{\alpha} = \frac{\theta_E^2}{\theta^2} \vec{\theta} \qquad (3.2)$$

with the Einstein radius for a point source defined as

$$\theta_E \equiv \sqrt{\frac{4MGD_{SL}}{D_S D_L c^2}}. \qquad (3.3)$$

Then the lens equation governing the positions of the images reduces to

$$\vec{\beta} = \vec{\theta} - \frac{\theta_E^2 \vec{\theta}}{\theta^2}. \qquad (3.4)$$

We can choose the coordinate so that the lens lies at the origin. Then, $\vec{\beta} = 0$ corresponds to the special case that the source is directly aligned with the lens. In that case, the lens equation is solved for all values of $\theta = \theta_E$. That is, an Einstein ring is observed around the lens at angular radius equal to θ_E.

If the source is not directly behind the lens, we can keep the lens at the origin and choose the coordinate system so that the source position lies along the x-axis and is positive: $\vec{\beta} = \beta\hat{x}$ with $\beta > 0$. Then the x- and y-components of the lens equation are

$$\beta = \theta_x \left[1 - \frac{\theta_E^2}{\theta^2}\right]$$

$$0 = \theta_y \left[1 - \frac{\theta_E^2}{\theta^2}\right]. \qquad (3.5)$$

Suppose that the position of the image has a nonzero y-component: $\theta_y \neq 0$. Then the second of these equations would require that the distance from the origin is $\theta^2 \equiv \theta_x^2 + \theta_y^2 = \theta_E^2$. But that would mean that the term in brackets in the first equation would vanish, so that the first equation could not be satisfied. We are forced to conclude that, in this simple point mass case, $\theta_y = 0$ so that the observed images are located in the 2D plane along a line connecting the lens and the true source position. It remains to for us to determine where along the line the images lie.

The positions of the images are completely determined by the x-component of the lens equation:

$$\beta = \theta - \frac{\theta_E^2}{\theta} \qquad (3.6)$$

where we can now drop the x subscript on θ because the vector lies only in the x-direction. This is a quadratic equation for the images, so it has two solutions:

$$\theta_{\pm} = \frac{\beta}{2}\left[1 \pm \sqrt{1 + \frac{4\theta_E^2}{\beta^2}}\right]. \qquad (3.7)$$

Fig. 3.2 shows the positions of the two images as the source position varies. The dashed curve traces the (unphysical) case that the observed image lies at the true position of the source. The two images are on either side of the lens with θ_x positive for one and negative for the other. When the source is close to the lens (β small compared to θ_E), the images are near the Einstein radius on either side. As the source moves farther away from the lens, the image on the other side of the lens moves towards the lens, while the primary image asymptotes closer and closer to the true source position.

It will be useful to record the two limits of Eq. (3.7) analytically. First, when β is small, the second term in the square root dominates so that

$$\theta_{\pm} \simeq \pm\theta_E + \frac{\beta}{2} \qquad (\beta \ll \theta_E). \qquad (3.8)$$

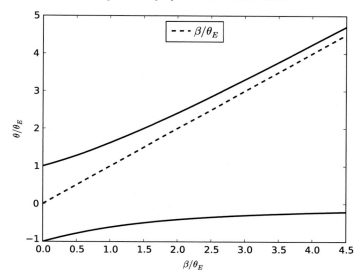

Figure 3.2 Images produced by a point mass lens. The lens is at $\theta = 0$ and the true source position is β; all angles are in units of the Einstein radius. The dashed curve is the observed position in the absence of lensing: $\theta = \beta$. In the presence of a lens, two images are observed, one on either side of the lens.

Then, when the source is far from the lens,

$$\theta_+ \simeq \beta + \frac{\theta_E^2}{\beta} \qquad (\beta \gg \theta_E)$$

$$\theta_- \simeq -\frac{\theta_E^2}{\beta} \qquad (\beta \gg \theta_E). \tag{3.9}$$

The first multiply imaged object discovered was the Twin Quasar, by Walsh, Carswell, and Weymann in 1979 using the 2.1 m telescope at the Kitt Peak Observatory. A more recent image of the Twin Quasar, taken by the Hubble Space Telescope, is shown in Fig. 3.3 and clearly shows two well-separated objects. Spectra of these two objects confirm that they are identical. The two images are separated by $6''$. Problem 3.1 offers an opportunity to estimate the mass of the lens.

3.2 Spherically Symmetric Distribution

The first obvious generalization of a point mass lens is to an extended lens that is spherically symmetric. We want to calculate the deflection angle generated by a spherically symmetric distribution. From Eq. (2.66), this is equal to

$$\vec{\alpha}(\vec{\theta}) = \frac{1}{\pi} \int d^2\theta' \, \kappa(\theta') \, \frac{\vec{\theta} - \vec{\theta}'}{|\vec{\theta} - \vec{\theta}'|^2} \tag{3.10}$$

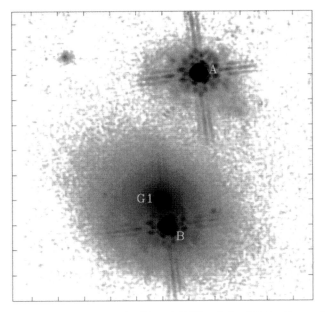

Figure 3.3 Two images (labelled "A" and "B") of the "Twin Quasar" from the Hubble Space Telescope (Keeton et al., 2000). The galaxy that serves as a lens is labelled "G1" and the separation of the two images is $6''$.

where κ is the surface density in units of the critical surface density Σ_{cr} as defined in Eq. (2.64). The symmetry here is encoded in the fact that κ depends only on the angular distance θ' and not on its direction. Looking at this integral, we see that it has a direction, i.e., it is a vector, and the only available vector in the problem (since the mass distribution has no preferred direction) is $\vec{\theta}$. Therefore, we can immediately write down that

$$\vec{\alpha}(\vec{\theta}) = A(\theta)\,\vec{\theta}. \tag{3.11}$$

The task becomes determining the value of the coefficient A, which of course will depend on the angular distance θ since the deflection will fall off far from the lens.

To determine A, we can integrate the divergence of $\vec{\alpha}$ over a disk centered on the lens and having a radius θ_{max}. Using Gauss's Theorem leads to

$$\int_{\theta < \theta_{max}} d^2\theta\, \vec{\nabla} \cdot \vec{\alpha}(\vec{\theta}) = \oint_C d\phi\, \vec{\theta} \cdot \vec{\alpha} \tag{3.12}$$

where the surface integral on the left extends out to θ_{max} and the contour integral on the right is around the edge of the disk, so the azimuthal angle ϕ extends from 0 to 2π. But that angular integral is trivial because the integrand is $A(\theta_{max})\theta_{max}^2$

and therefore does not depend on ϕ; it depends only on the angular distance. The right-hand side of Eq. (3.12) is then simply $2\pi A(\theta_{max})\theta_{max}^2$.

The left-hand side of Eq. (3.12) is a little tricky so let's walk through the calculation slowly. The integrand is

$$\vec{\nabla} \cdot \vec{\alpha}(\vec{\theta}) = \frac{1}{\pi} \int d^2\theta' \, \kappa(\theta') \, \vec{\nabla}_{\vec{\theta}} \cdot \frac{\vec{\theta} - \vec{\theta}'}{|\vec{\theta} - \vec{\theta}'|^2} \tag{3.13}$$

where the subscript on $\vec{\nabla}_{\vec{\theta}}$ instructs us to take the divergence with respect to $\vec{\theta}$, not $\vec{\theta}'$. The problem boils down to taking the divergence of the vector $(\vec{\theta} - \vec{\theta}')/|\vec{\theta} - \vec{\theta}'|^2$. Acting on the numerator and the denominator with the divergence leads to the apparent conclusion that the divergence vanishes. The sticking point is that there is a singularity at the point $\vec{\theta} = \vec{\theta}'$, so the divergence there is proportional to a 2D Dirac delta function. Specifically,

$$\vec{\nabla}_{\vec{\theta}} \cdot \frac{\vec{\theta} - \vec{\theta}'}{|\vec{\theta} - \vec{\theta}'|^2} = 2\pi \delta_D^2(\vec{\theta} - \vec{\theta}'), \tag{3.14}$$

a result that can be proved in Exercise (3.2).

Armed, then, with the identity of Eq. (3.14), we find that

$$\vec{\nabla} \cdot \vec{\alpha} = 2\kappa(\theta) \tag{3.15}$$

so now equating the left- and right-hand sides of Eq. (3.12) leads to

$$2 \int_{\theta < \theta_{max}} d^2\theta \, \kappa(\theta) = 2\pi A\theta_{max}^2. \tag{3.16}$$

This fixes the coefficient $A(\theta)$ to be

$$A(\theta) = \bar{\kappa}(\theta) \tag{3.17}$$

where the mean value of the normalized surface density is defined as

$$\bar{\kappa}(\theta) \equiv \frac{1}{\pi \theta^2} \int_{\theta' < \theta} d^2\theta' \, \kappa(\theta'). \tag{3.18}$$

So, we arrive at the lens equation for a spherically symmetric mass distribution:

$$\vec{\beta} = \vec{\theta} - \bar{\kappa}(\theta)\vec{\theta}. \tag{3.19}$$

Recall that $\kappa(\theta)$ is the surface density an angular distance θ from the lens divided by the critical surface density. So, the lens equation dictates that the deflection (the difference between the true source position $\vec{\beta}$ and the observed image $\vec{\theta}$) an angular distance θ from the lens is governed by the mass contained within the cylinder of radius $D_L\theta$. Note that this is a little bit different from the total mass within a 3D radius $D_L\theta$ (see Exercise (3.3)).

Let's work through this in equations. The mean value of κ is

$$\frac{1}{\pi \theta^2} \int_{\theta' < \theta} d^2\theta' \, \kappa(\theta') = \frac{1}{\pi D_L^2 \theta^2} \int_{R' < R} d^2R' \, \frac{\Sigma(\vec{R'})}{\Sigma_{cr}} \tag{3.20}$$

where the relation between the transverse distance from the lens $R = \theta D_L$ has been used. The surface density is the integral of the density all along the line of sight, so the integral d^2R' of the surface density over the disc of radius R results in the total mass contained within a cylinder with radius R:

$$M(R) \equiv \int_{R' < R} d^2R' \, \Sigma(R'). \tag{3.21}$$

In terms of this mass,

$$\bar{\kappa}(\theta) = \frac{M(R = D_L\theta)}{\pi D_L^2 \theta^2 \Sigma_{cr}}. \tag{3.22}$$

This is sometimes written more concisely by recognizing that $M(R)/\pi R^2$ is the mean surface density within R, call it $\bar{\Sigma}(R)$. Then the deflection angle is simply $(\bar{\Sigma}(R)/\Sigma_{cr}) \, \vec{R}/D_L$.

3.3 Isothermal Sphere

We now apply this result to the case where the distribution takes a particularly simple form, that of an isothermal sphere. The density in this case is

$$\rho(r) = \frac{\sigma^2}{2\pi G(r^2 + r_c^2)} \tag{3.23}$$

where r_c is called the core radius, the radius within which the density turns flat, and σ is the velocity dispersion of the elements that comprise the distribution (stars in a galaxy or galaxies in a galaxy cluster), which also serves as an indicator of the mass of the distribution.

To get a sense of the relation between velocity dispersion and mass, consider Fig. 3.4, which shows the velocity dispersion of galaxies in galaxy clusters. The velocity dispersions of 43 galaxy clusters are shown as a function of their masses, determined independently using a variety of probes. For us, the take-away is that velocity dispersions σ of order 1000 km/sec, are associated with masses of order $5 \times 10^{14} M_\odot$. Exercise (3.4) provides some detail. You can show there that mass, suitably defined, scales as σ^4, so when we consider a galaxy with a mass of order $10^{12} \, M_\odot$, we should have in mind values of $\sigma \simeq 200$ km/sec.

The isothermal profile in Eq. (3.23) becomes constant ($\rho \propto r^0$) at small distances from the center and falls off as r^{-2} at large distances. Both of these limits are a bit off: numerical simulations indicate that profiles of gravitationally bound objects

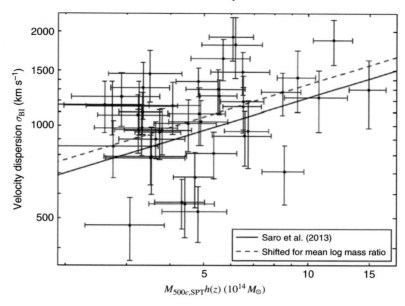

Figure 3.4 Velocity dispersion of galaxy clusters as a function of their masses, from Ruel et al. (2014). The masses have been inferred independently using information besides the velocity dispersion. The label on the x-axis M_{500c} reflects the fact that, for extended objects like galaxy clusters, mass needs to be clearly defined as that within a given radius (see Exercise (3.4)).

like galaxy clusters tend to rise towards the center and fall off faster than r^{-2} moving away, but the isothermal profile remains instructive because it both allows for easy analytic solutions and is hard to distinguish observationally from more accurate forms.

Given the density profile in Eq. (3.23), the first step is to compute the surface density

$$\Sigma(R) = \frac{\sigma^2}{2\pi G} \int_{-\infty}^{\infty} \frac{dz}{R^2 + z^2 + r_c^2}. \qquad (3.24)$$

To carry out the integral, note that it is symmetric in $z \to -z$ so we can limit the range to positive z and multiply by 2. Then define a new dummy variable $\tan\theta \equiv z/(R^2 + r_c^2)^{1/2}$, so that $d\theta = \sqrt{R^2 + r_c^2}\, dz/(z^2 + R^2 + r_c^2)$; i.e., the very integrand of interest multiplied by a constant. As z runs from 0 to ∞, θ runs from 0 to $\pi/2$. So,

$$\Sigma(R) = 2 \times \frac{\sigma^2}{2\pi G \sqrt{R^2 + r_c^2}} \int_0^{\pi/2} d\theta$$

$$= \frac{\sigma^2}{2G \sqrt{R^2 + r_c^2}}. \qquad (3.25)$$

The next step is to compute the mean surface density within a given radius:

$$M(R) = 2\pi \int_0^R dR' R' \Sigma(R')$$

$$= \frac{\pi \sigma^2}{G} \left[\sqrt{R^2 + r_c^2} - r_c \right]. \tag{3.26}$$

So the lens equation for an isothermal sphere becomes

$$\vec{\beta} = \vec{\theta} - \frac{\vec{\theta}}{\theta^2} \frac{\sigma^2}{G D_L \Sigma_{\rm cr}} \left[\sqrt{\theta^2 + \theta_c^2} - \theta_c \right] \tag{3.27}$$

where all distances from the center have been expressed as angular distances with the usual $\theta = R/D_L$ supplemented by the definition of the core angular radius $\theta_c \equiv r_c/D_L$.

We can define the Einstein radius of an isothermal sphere as:

$$\theta_0 \equiv \frac{4\pi \sigma^2 D_{SL}}{D_S c^2}. \tag{3.28}$$

To give a sense of scale, when the source is far behind the lens the distance ratio in Eq. (3.28) drops out. Then, using $\sigma \simeq 1000$ km/s appropriate for a massive galaxy cluster, θ_0 is of order $0.5'$. If the lens is a galaxy with a velocity dispersion smaller by a factor of 4, then $\theta_0 \simeq 2''$. Even the smaller of these is well within reach of even ground-based telescopes, which can achieve resolution greater than $1''$. In terms of this effective Einstein radius, the lens equation for the isothermal sphere is

$$\vec{\beta} = \vec{\theta} - \frac{\vec{\theta} \theta_0}{\theta^2} \left[\sqrt{\theta^2 + \theta_c^2} - \theta_c \right]. \tag{3.29}$$

3.3.1 Singular Isothermal Sphere

When the mass distribution has no core, $\theta_c = 0$, the lens equation reduces to

$$\vec{\beta} = \vec{\theta} \left[1 - \frac{\theta_0}{|\theta|} \right]. \tag{3.30}$$

If the source is aligned directly behind this lens, then the background point will be imaged on to a ring a distance θ_0 from the cluster center. Indeed, this motivates the definition of θ_0 as the Einstein radius for an isothermal sphere.

If the source is off-center, $\beta \neq 0$, the situation is similar to the point mass: the image $\vec{\theta}$ has a component only along the axis defined by the line connecting the true source position to the lens center, $\vec{\beta}$. The image(s) along this line have position determined by

$$\beta = \theta - {\rm sign}(\theta)\theta_0. \tag{3.31}$$

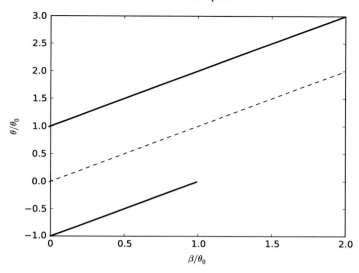

Figure 3.5 Images in the presence of an isothermal sphere with no core, with all angles in units of the effective Einstein angle θ_0, defined in Eq. (3.28). Upper line shows one image, always displaced from the true position by an amount equal to θ_0. The second image appears on the other side of the lens (located at the origin), but only if the true source position is within θ_0. Dashed line shows the image in the absence of lensing.

For positive θ, this means the image will be at $\theta_+ = \beta + \theta_0$, i.e., farther from the lens than is the true source, as indicated in Fig. 3.5. Negative θ is also allowed – another image on the other side of the lens. This second image is at position $\theta_- = \beta - \theta_0$. For consistency, though (since this solution must be negative), the second image exists only if β is smaller than θ_0. To sum up, a source far from the projected center of the lens produces a single image slightly displaced, at $\theta_+ = \beta + \theta_0$. As the source gets closer to the center and comes within a distance θ_0, another image appears, initially at $\theta_- = 0$ but moving outwards until it reaches $-\theta_0$ when the source is completely aligned with the lens. An oddity of this artificial profile is that the difference between the lens-side image θ_+ and the true source position is constant. This is a bit counterintuitive: shouldn't the effect of lensing get smaller as one moves away from the lens? The feature here is solely an artifact of the assumption that the density falls as θ^2 so that the mass enclosed within a given angular radius continues to rise as one moves away from the lens.

3.3.2 Cored Isothermal Sphere

The mass density in galaxies and in galaxy clusters does not continue to rise as r^{-2} all the way into the center, so keeping a nonzero core radius is a much better approximation to reality than is the singular isothermal profile. In that case, we need to consider the full lens equation (3.29). As in the singular case, when the source is perfectly aligned with the lens ($\beta = 0$), the image is a ring, this time at a radius determined by

$$\theta^2 = \theta_0 \left[\sqrt{\theta^2 + \theta_c^2} - \theta_c \right]. \tag{3.32}$$

Solving for θ leads to an expression for the Einstein radius of a cored isothermal sphere:

$$\theta_E = \theta_0 \sqrt{1 - 2\frac{\theta_c}{\theta_0}}. \tag{3.33}$$

So the Einstein radius changes a bit from the uncored result. If the core radius is too large – $\theta_c > \theta_0/2$ – there is no solution. More generally, a small core will reduce the radius of the ring.

It is useful to estimate the value of the core radius at which the Einstein ring ceases to exist; i.e., at what radius does the density profile need to flatten out so that there will not be a ring even if a source is directly aligned with the center of the lens. When the lens is a galaxy, we estimated above that θ_0 is of order a few arcseconds, so that a core larger than $1''$ would not produce a ring. A typical order-of-magnitude distance for a cosmological object is 1 Gpc $= 10^3$ Mpc. The physical size of $1''$ at this distance is then about 5 kpc. The transition point – the point at which the slope of the profile changes to be flatter – is typically quite a bit larger than this for galaxies. For example, for the Milky Way, the best guess is of order 20 kpc. So the core radius is likely to be too large in most galaxies for an Einstein ring to be produced. This is not a theorem: as the galaxies get farther from us, a one arcsecond core radius corresponds to a *larger* physical size, so there are likely to be more galaxies that can produce rings (or giant arcs) as one observes deeper. For galaxy clusters, with their much larger values of θ_0, the critical size of a core is roughly a factor of 10–20 times larger, or hundreds of kpc. This is quite a bit larger than the cores in most clusters, so clusters are unlikely to produce dramatic signatures such as Einstein rings.

Fig. 3.6 shows a dramatic example, the Einstein ring formed when a galaxy 950 Mpc away from us serves as a lens for a much more distant galaxy, called SDP.81, 1.6 Gpc from us. The velocity dispersion of the lens is 265 km/sec, so $\theta_0 = 1.4''$. The core, on the other hand, has been determined via high-resolution radio images to be much smaller than this: $\theta_c \sim 0.15''$. So a brilliant Einstein ring is produced.

If the source is not directly behind the lens, then the lens equation becomes more complicated. It still has the property that the image(s) lie(s) along the line connecting the lens center to the source β, so we can again choose $\vec{\beta}$ and $\vec{\theta}$ to lie

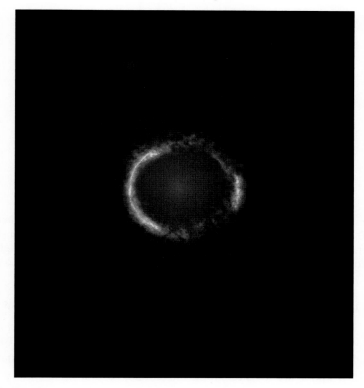

Figure 3.6 Einstein ring formed when a distant background galaxy, SDP.81, is lensed by a foreground galaxy. The ring has a radius of about $1''$. The lensing galaxy can be observed in the center; it has a core of about $0.15''$ so Eq. (3.33) still allows for the existence of an Einstein ring. Credit: ALMA (NRAO/ESO/NAOJ); B. Saxton NRAO/AUI/NSF; NASA/ESA Hubble, T. Hunter (NRAO).

along the x-axis: $\vec{\beta} = (\beta, 0)$ and similarly for θ. Finally, we can again choose the x-axis so that β is positive. Then the lens equation (3.29) reduces to

$$\theta \, (\beta - \theta) = -\theta_0 \left[\sqrt{\theta^2 + \theta_c^2} - \theta_c \right]. \tag{3.34}$$

If we think of solving this equation for the images θ as a function of β, the problem seems quite complex. However, a much simpler thing to do is to plot β as a function of θ and then reverse the x- and y-axes. This is shown in Fig. 3.7 for a particular value of the core radius. For large values of source position β, there is a single image located at $\theta \sim \beta + \theta_0$, just as in the case of the singular isothermal sphere. As β moves closer to the lens, two new images appear, gradually separating as β gets smaller. When the source is perfectly aligned with the center of the lens, one image migrates to the center and the other two form part of the Einstein ring with radius given by Eq. (3.33). Our goal in the next section is to understand this

Multiple Images

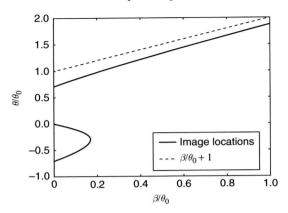

Figure 3.7 Locations of the images produced by a cored isothermal sphere with $\theta_c = 0.25\theta_0$. Dashed line shows the position of the image produced by a singular isothermal sphere. Three images are produced if the source is close enough to the lens; for this value of θ_c, this translates into the requirement that $\beta < 0.2\theta_0$.

behavior, in particular the special value of β within which a new pair of images forms.

Before turning to that question, though, it is worth thinking more broadly about this special value of β. We have been working with the one-dimensional equation, where both $\vec{\beta}$ and $\vec{\theta}$ are aligned along the x-axis, so that β_c is simply a point on the line. More generally, as depicted in Fig. 3.8, the critical value of β here will be the radius of a circle, so all values of β that fall within the circle will produce multiple images, while those outside will produce only a single image. The circle with radius β_c is called a *caustic* curve, so our next goal is to determine the caustic curve of a cored isothermal sphere.

3.4 Caustics for the Cored Isothermal Sphere

Our aim is to determine the radius of the circle that delineates between regions in the source plane that produce one image and that produce multiple images. Let's return to the lens equation and write it as

$$\beta(\theta) = \theta - \frac{\theta_0}{\theta}\left[\sqrt{\theta^2 + \theta_c^2} - \theta_c\right]. \tag{3.35}$$

The function $\beta(\theta)$ goes to $\pm\infty$ as θ goes to $\pm\infty$, so it crosses a particular value of β at either one or three places, as shown in Fig. 3.9. In order for there to be three solutions to this equation, the function must have two local extrema at $\theta_{\pm}$.

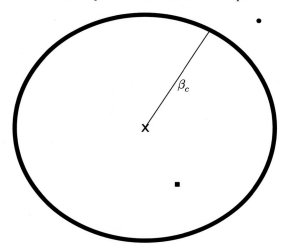

Figure 3.8 For an isothermal sphere, the circle in the source plane with radius β_c that delineates between regions where a source will produce three images (e.g., the filled square) and those where a source will produce only a single image (e.g., the filled circle).

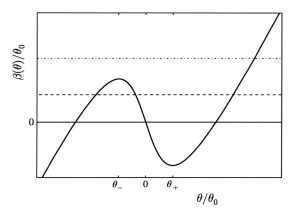

Figure 3.9 Generic plot of $\beta(\theta)$ for an isothermal sphere, from Eq. (3.35). If the source position β is at the position of the dashed line, then there are three values of θ that solve the lens equation since the polynomial crosses that value of β three times. If β is further away, at the position of the dot-dashed line, then there will be only one image.

Once we find the extrema depicted in Fig. 3.9, then any value of β smaller than $\beta(\theta_-)$ will produce three images (recall that, by convention, β is positive). To see whether these extrema exist, and to find their values, we set the derivative to zero:

$$\frac{d\beta}{d\theta} = \left(1 - \frac{\theta_0 \theta_c}{\theta^2}\right) + \frac{\theta_0 \theta_c^2}{\theta^2 \sqrt{\theta^2 + \theta_c^2}} = 0. \tag{3.36}$$

The second term is positive here, so the sum of the two terms can be zero only when the first is negative or

$$|\theta| < \sqrt{\theta_0 \theta_c}.\tag{3.37}$$

Remembering this constraint, then isolating the square root in Eq. (3.36), and squaring both sides leads to a quadratic equation for $\theta^2/\theta_0\theta_c$ that determines the location of the extrema:

$$\left(\frac{\theta^2}{\theta_0\theta_c}\right)^2 - \frac{\theta^2}{\theta_0\theta_c}\left[2 - \frac{\theta_c}{\theta_0}\right] + 1 - 2\frac{\theta_c}{\theta_0} = 0.\tag{3.38}$$

The two solutions to this quadratic equation are

$$\frac{\theta^2}{\theta_0\theta_c} = 1 + \frac{\theta_c}{2\theta_0}\left[\pm\sqrt{1 + \frac{4\theta_0}{\theta_c}} - 1\right],\tag{3.39}$$

but if we choose the plus sign, the term in square brackets is greater than zero. Then the right-hand side will be greater than one, so θ^2 would be greater than $\theta_0\theta_c$, which violates the condition in Eq. (3.37). Therefore, the relevant solution is the one with the minus sign; taking the square root of this solution leads to extrema at

$$\theta_\pm = \pm\sqrt{\theta_0\theta_c}\left(1 - \frac{\theta_c}{2\theta_0}\left[\sqrt{1 + \frac{4\theta_0}{\theta_c}} + 1\right]\right)^{1/2}.\tag{3.40}$$

The term in parentheses needs to be positive, which corresponds to the constraint we've already encountered when considering an Einstein ring: a cored isothermal profile can produce multiple images only if $\theta_c < \theta_0/2$.

We can now get more insight as to when the multiple images appear by plugging[1] Eq. (3.40) into Eq. (3.35) to obtain the caustic radius $\beta_c \equiv \beta(\theta_-)$. Fig. 3.10 shows this radius as a function of the core radius θ_c. For a particular value of the core radius, all values of β smaller than β_c will produce multiple images. To get a feel for the numbers, if a galaxy has $\theta_c = 1''$ and $\theta_0 = 3''$, then the caustic would lie at $\beta_c = 0.3''$. That is, the alignment between background source and the foreground lens would have to be pretty tight.

We have identified a caustic in the source plane, that is a special value of β, such that there is a change in the number of images as β crosses this special value. That special value is β at the locations where

$$\frac{d\beta}{d\theta} = 0.\tag{3.41}$$

For the isothermal sphere, this is a circle in the source plane: if the source is contained within this circle, there will be three images; if it is outside the circle,

---

[1] We will use $\beta(\theta_-)$, but $\beta(\theta_+)$ would simply be $-\beta(\theta_-)$, so their magnitudes are the same (see Exercise (3.9)), and either can be used to obtain the caustic radius β_c.

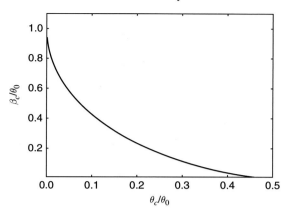

Figure 3.10 As a function of the core radius of an isothermal sphere, the value of β that satisfies $d\beta/d\theta = 0$: the caustic. Values of β below the curve lead to multiple images; above, only a single image.

there will be only one image. The caustic curve in this case is a circle; more generally, when the mass distribution of the lens is not spherically symmetric, the caustic curves will have more structure.[2] Physically, the vanishing derivative means that the source position changes hardly at all as the image positions change, or equivalently, the image positions undergo large shifts for a very small shift in the source position near the caustic. We will see in the next chapter that caustics carry even more significance as we piece together the total phenomenon of gravitational lensing.

Finally, consider Fig. 3.11, which shows the familiar circular caustic of an isothermal sphere. This curve is in the source plane, as it is the relevant curve for the source position $\vec{\beta}$. The two other curves in the figure are *critical curves*; they illustrate the location of the images when the source lies on the caustic. The critical curves therefore lie in the image plane.

3.5 Time Delays

The time it takes light to reach us from a source will be different for rays taking different paths, so flux emitted from a multiply imaged object will reach us at different times. There are two contributions to this delay: the obvious geometric delay since the path lengths differ and the Shapiro time delay that we have already encountered. So,

$$\Delta t = \Delta t_{\text{Geom}} + \Delta t_{\text{Sh}}. \tag{3.42}$$

---

[2] In these more general cases, we will see that the criterion for a caustic changes from Eq. (3.41) to the two-dimensional analog: the determinant of the 2×2 matrix $\partial\beta_i/\partial\theta_j$ vanishes at the caustics. Even in the spherically symmetric case treated here, this generalization picks up a second caustic at the origin (see Exercise (3.10)).

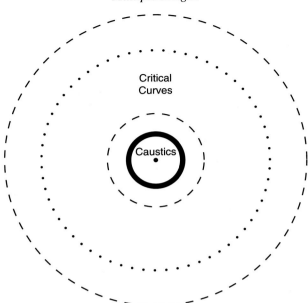

Figure 3.11 Solid circle shows the caustic for an isothermal sphere, while the two dashed circles are the critical curves, the locations of the images produced when a source lies on the caustic. The dot at the center is a second caustic, with the Einstein radius (dotted circle) the corresponding critical curve (see Exercise (3.10)).

The effect of the Shapiro time delay is best captured from Eq. (2.42). There, replacing $-MG_{\odot}/r$ by the 3D gravitational potential ϕ, we see that the Shapiro time delay is the integrated gravitational potential. So,

$$\Delta t_{\text{Sh}} = \frac{-2}{c^3} \int dz \, \phi(z)$$

$$= -\frac{\Phi}{c^2} \times \frac{D_S D_L}{D_{SL} c} \tag{3.43}$$

where the second line comes from removing the distances from the definition of the projected potential in Eq. (2.55). Although the dependence has been suppressed, Φ depends upon the incoming direction $\vec{\theta}$ so the time delay also depends on the direction and will typically differ from one image to the next.

The geometric effect requires us to compute the difference in path length between the undeflected ray (heavy dashed in Fig. 3.12) and the deflected ray (heavy solid). The length of the undeflected path in Fig. 3.12 is easily seen to be equal to

$$D_u = \frac{D_S}{\cos \beta}. \tag{3.44}$$

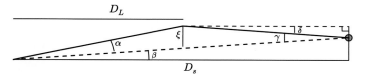

Figure 3.12 Geometric time delay. The two angles α and γ are related because the perpendicular distance ξ is equal to both $D_L\alpha$ and $D_{SL}\gamma$, a relation that can be used to determine the difference in the unlensed path length (heavy dashed) and lensed path length (heavy solid).

The deflected path is composed of two parts, so that

$$D_d = \frac{D_L}{\cos(\alpha + \beta)} + \frac{D_{SL}}{\cos\delta} \tag{3.45}$$

where α and β are the standard angles (deflection and source) and δ is shown in Fig. 3.12. There are three angles that combine to give the ($180°$) line on the source plane: γ in the middle, the angle complementary to δ on one side (it plus δ add up to $90°$), and the angle complementary to β on the other side. Therefore, $\gamma + (\pi/2 - \delta) + (\pi/2 - \beta) = \pi$ or

$$\delta = \gamma - \beta. \tag{3.46}$$

But, as described in the caption, $\gamma = D_L\alpha/D_{SL}$, so

$$\delta = \alpha\frac{D_L}{D_{SL}} - \beta. \tag{3.47}$$

We can now Taylor expand the cosines in the denominators of Eq. (3.45) to obtain an expression for the path length difference:

$$\begin{aligned} D_d - D_u &= \frac{D_L(\alpha + \beta)^2}{2} + \frac{D_{SL}(D_L\alpha/D_{SL} - \beta)^2}{2} - \frac{D_S\beta^2}{2} \\ &= \frac{D_L D_S}{2D_{SL}}\left(\vec{\theta} - \vec{\beta}\right)^2 \end{aligned} \tag{3.48}$$

remembering that geometrically $\vec{\alpha} \equiv \vec{\theta} - \vec{\beta}$. Dividing by c gives the geometric time delay, so that the full time delay is equal to

$$\Delta t = (1 + z_L)\frac{D_L D_S}{cD_{SL}}\left[\frac{\left(\vec{\theta} - \vec{\beta}\right)^2}{2} - \frac{\Phi}{c^2}\right]. \tag{3.49}$$

The factor of $1 + z_L$, where z_L is the *redshift* of the lens, in front has been added to keep the formula accurate even for cosmological applications. The basic idea is that light emitted when the universe was younger will be delayed in its journey as the universe expands, and the factor of $1 + z_L$ accounts for this delay.

There are two elegant features of Eq. (3.49). First, note that the prefactor is a ratio of distances, having nothing to do with the details of the lens, while the terms in square brackets depend only on the mass distribution of the lens. So the problem breaks up into two parts, and if the lens modelers do their jobs, then cartographers interested in mapping distances to objects can obtain distances from time delays. The second feature of Eq. (3.49) is that it leads immediately to the lens equation. If we insist on minimizing the time traveled by light rays – i.e., if we invoke *Fermat's principle* – then differentiating Eq. (3.49) with respect to angle results in:

$$\frac{d}{d\theta^i} \left[\frac{(\vec{\beta} - \vec{\theta})^2}{2} - \frac{\Phi}{c^2} \right] = 0, \tag{3.50}$$

which immediately reduces to the lens equation since $\vec{\alpha} = \nabla\Phi/c^2$. So the lens equation can be seen as a manifestation of Fermat's principle.

Let us now turn to how this can be detected. Imagine a source that emits a pulse of light, and that light reaches us – due to an intervening lens – from multiple directions. The time difference between the blips arriving from two different directions will be determined by the lens modeling factor in square brackets in Eq. (3.49):

$$t_1 - t_2 = (1 + z_L) \frac{D_L D_S}{c D_{SL}} \left(\left[\frac{(\vec{\theta}_1 - \vec{\beta})^2}{2} - \frac{\Phi(\theta_1)}{c^2} \right] - \left[\frac{(\vec{\theta}_2 - \vec{\beta})^2}{2} - \frac{\Phi(\theta_2)}{c^2} \right] \right). \tag{3.51}$$

Consider the point mass case with β much smaller than the Einstein radius. Then, the two images form at the positions θ_+ given in Eq. (3.8) and the differences $(\theta_+ - \beta)^2$ and $(\theta_- - \beta)^2$ are nearly equal so the geometric part of the time delay is insignificant. The main contribution in this case comes from the difference in the potentials. Using Eq. (2.67), we find

$$\Phi(\theta_+) - \Phi(\theta_-) = c^2 \theta_E^2 \ln \left| \frac{\theta_+}{\theta_-} \right| \tag{3.52}$$

In the limit of small β, the logarithm can be Taylor expanded so that

$$\Phi(\theta_+) - \Phi(\theta_-) \simeq 2c^2 \theta_E \beta. \tag{3.53}$$

So for the point mass, when the source is close to the z-axis, the time difference between the two images is

$$t_+ - t_- \Big|_{\text{point mass}} \simeq -(1 + z_L) \frac{D_L D_S}{c D_{SL}} 2\theta_E \beta. \tag{3.54}$$

Several points are in order here. First note that the sign is correct: the light that passes closest to the lens (that seen at θ_-) is delayed more than that which travels further from the lens. Therefore the θ_+ image will arrive earlier; hence the minus sign. Let's obtain an order of magnitude estimate for the amount of the delay. If the distances are cosmological, the entire prefactor $(1 + z_L)D_L D_S/D_{SL}$ is of order 1 Gpc. Then taking the lens to be a galaxy with mass $10^{12} M_\odot$ a distance 1 Gpc away leads to an estimate of one and a half months if $\beta = 0.1\theta_E$. But this delay is very sensitive to β. You can show this in Exercise (3.12): the delay gets as large as three years when $\beta = 2\theta_E$. Indeed, we will shortly see that this wide variation extends to observed delays.

The above point mass example is very simplistic but it offers a nice segue into how time delays can determine the distance ratio $D_L D_S/D_{SL}$. In this very simple case, there are three observables: the time delay, and the angular distances of the two images from the lens. There are fortuitously also three unknowns in the model: the true source position β, the Einstein radius θ_E, and the distance ratio $D_L D_S/D_{SL}$. Therefore, an analyst can determine the distance ratio from the observations.

Fig. 3.13 shows an example of a strongly lensed system for which time delays have been measured. The background source is a galaxy, inside of which resides a compact Active Galactic Nucleus (AGN), which emits a time-varying flux. There are four images of this AGN apparent in the figure. The extended galaxy is also apparent in the figure: it appears as the extended arcs that virtually form an Einstein ring. The foreground lenses are a galaxy labeled 'G' and a small satellite 'S'. The time variation of the AGN is useful because we can measure the time difference between the flux maxima, for example. Sophisticated analysis of over nine years of data has determined that the flux from A, B, and C arrives roughly at the same

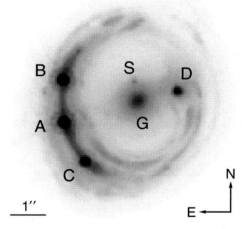

Figure 3.13 High-resolution image from the Hubble Space Telescope of the gravitational lens RXJ1131-1231 (Suyu et al., 2013). Four points labeled A,B,C,D are images of the same Active Galactic Nucleus. The main lensing galaxy is denoted G, while a nearby smaller satellite is S.

time (within a day), but the D image is delayed by 90 days. This is qualitatively consistent with our point mass example, where the image closest to the lens was delayed with respect to the other.

Extracting the distance ratio from this system also follows qualitatively our point mass discussion. In that case, there were three observables and three unknown parameters. In this more complex case, Suyu et al. (2013) used as input the flux in over 25,000 pixels that covered the region shown in Fig. 3.13. With these 25,000 observables, they were able to constrain 39 parameters that described the lens and the source. They concluded by inferring the distance ratio $D_{SL}/D_L D_S$ to about 10 percent accuracy.

This level of complexity and range of delays lead us back to the remarkable lensed supernova depicted in Fig. 3.1. There were no delays observed between any of the four images shown there. However, modelers of the system determined that there are likely to be two additional images of the supernova since there are two additional images of the host galaxy. The initial four images were detected in April 2015; Fig. 3.14 shows predictions from different models of the lens for when the next supernova image was supposed to appear. Indeed, the predicted fifth image did appear at the end of 2015.

Suggested Reading

The initial discovery paper of the Twin Quasar (Walsh et al., 1979) is refreshingly short and of course ground-breaking. I particularly enjoyed this understated passage from the conclusion: "... we may be dealing with a single source which has

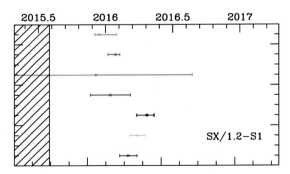

Figure 3.14 Predicted date of the fifth image of the lensed supernova shown in Fig. 3.1 (Treu et al., 2016). The different bars represent different models of the lens mass distribution. The label 'SX/1.2 -S1' reflects the fact that the first image of the supernova is dubbed S1; there is an image of the host galaxy wherein the supernova was not observed in 2015 (that galaxy is '1.2'), and the expected super- nova image is called 'SX'. So the bars show the predictions for the time at which the supernova image SX will appear in galaxy 1.2.

been split into two images by a gravitational lens. We shall consider this possibility after examining the more conventional explanation involving two distinct QSOs."
There have been many detailed studies of the lens(es) that produce this particular double image. A comprehensive one to look at is Keeton et al. (2000), which demonstrates why the estimate in Exercise (3.1) is too naive (although, in our defense, they use much more information than is contained in the exercise).

It is also interesting to trace the multiple image references backwards. The 1979 discovery paper gives a single reference for gravitational lensing, a 1971 paper by Sanitt, which in turn lists as a first reference a one-column paper by Zwicky (1937). It opens by mentioning that Einstein, as we have seen, had proposed that stars could serve as gravitational lenses. The best part of the Zwicky paper, though, is the last paragraph, which concludes by first reminding the reader that his measurements of velocities suggested that galaxy clusters (or "nebulae") were more massive than expected and then with the uncanny prediction that "Observations on the deflection of light around nebulae may provide the most direct determination of nebular masses and clear up the above-mentioned discrepancy."

A very clear and elegant article describing time delays is Blandford and Narayan (1986). As observations have improved, optimism has risen that time delays will emerge as a powerful method of determining cosmological distances. Some nice examples of the detailed modeling required are provided in Fadely et al. (2010) and Suyu et al. (2013). The number of events that might be observed in large surveys is projected in Oguri and Marshall (2010), while the resulting cosmological implications are described in Linder (2011).

Exercises

3.1   The two images of the Twin Quasar in Fig. 3.3 are $6''$ apart. The source quasar is 1.8 Gpc away from us, while the lens galaxy, labelled in the figure as "G", is 1.1 Gpc away from us. From the angular separation of the two images, estimate the Einstein radius of the lens, and from that, obtain an estimate of the mass of the lensing galaxy.

3.2   Prove the identity in Eq. (3.14). An easy way to see this is to set $\vec{\theta}' = 0$ and integrate both sides over all $\vec{\theta}$ out to some distant angular radius θ_{max}. The term on the right trivially reduces to 2π, while the one on the left can again be treated with Gauss's Theorem.

3.3   Consider a mass distribution that falls off as r^2. The mean value of κ involves an integral over the cylinder centered on the lens extending out to a given radius:

$$\bar{M}(R) \equiv \int_{R' < R} d^2 R' \int_{-\infty}^{\infty} dz\, \rho \left(\sqrt{R'^2 + z^2} \right). \qquad (3.55)$$

Compare this mass to the mass enclosed within a sphere of radius R for this mass distribution.

3.4   The mass of an isothermal sphere must be defined carefully, because at face value the total mass diverges if the profile is continued indefinitely: $\int d^3r/r^2$ diverges at large r. A conventional way to do this is to compute the mass contained within a given radius. Then, the question of mass definition reduces to the question of which radius to choose. A common one is r_{200}, the radius within which the mass density is equal to 200 times a canonical density known as the *critical density*.[3] So, this radius is determined by the equation

$$\frac{m_{200}}{4\pi r_{200}^3/3} = 200\rho_c. \tag{3.56}$$

The left-hand side is the mass within this radius, called m_{200}, divided by the volume of a sphere with this radius, so it is indeed the mean density within r_{200}. Let's postpone a discussion of the definition of ρ_c and note that m_{200} depends on r_{200} because

$$m_{200} = \int_{r<r_{200}} d^3r\rho(r). \tag{3.57}$$

Determine the relation between m_{200} and r_{200} for the singular isothermal sphere. The core radius is often much smaller than r_{200}, so setting r_c to zero is often a good approximation when defining this relation. Use the relation and the definition in Eq. (3.56) to obtain a relation for the mass as a function of the critical density and velocity dispersion.

3.5   For the isothermal sphere with dispersion $\sigma = 1000$ km/s, compute r_{200}, the radius within which the average density is equal to 200 times the critical density of the universe. What is the mass enclosed within this radius, r_{200}? Take the critical density to be $\rho_c = 10^{-29}$ g/cm^3.

3.6   Determine $\vec{\alpha}$ (numerically if necessary) for a cluster with profile

$$\rho(r) = \frac{\rho_0}{(r/r_s)(r+r_s)^2}. \tag{3.58}$$

3.7   Plot β against θ (both in units of θ_0) using Eq. (3.34) for $\theta_c = 0.2\theta_0$ and $\theta_c = 0.6\theta_0$. Reverse the axis so θ appears on the y-axis. Determine the value of β_c you get in the first case, the value of β at which two new images appear. Compare your result with an estimate gleaned from Fig. 3.10.

3.8   We showed that the Einstein radius of the isothermal sphere got smaller when θ_c moved away from zero. Of course, the total mass of the lens also

---

[3]   This critical density is not to be confused with the critical surface density, Σ_{cr}, that enters into the deflection by a single lens. Rather, as we will see in later chapters, it is related to the overall mass density in the universe.

gets smaller in that case. What happens if the total mass is kept fixed as θ_c changes? Compute the mass M within a sphere of radius θ_0, as defined in Eq. (3.28), for a cored isothermal mass distribution. This should depend on both θ_0 and θ_c. Show how θ_0 must be changed as θ_c is changed if M is to be kept fixed. Using this constraint, determine the change in the Einstein radius $\theta_0\sqrt{1 - 2\theta_c/\theta_0}$ as θ_c increases. Does this Einstein radius get larger or smaller?

3.9 Show that the lens equation for a cored isothermal sphere remains unchanged under the transformation

$$\vec{\theta} \to -\vec{\theta} \qquad \text{AND} \qquad \vec{\beta} \to -\vec{\beta}. \tag{3.59}$$

Use this fact to prove that $\beta(\theta_-) = -\beta(\theta_+)$ where $\theta_\pm$ are determined in Eq. (3.40).

3.10 As indicated in the footnote on page 53, and as we will derive in subsequent chapters, the true definition of a caustic is when the determinant of the matrix $\partial\beta_i/\partial\theta_j$ vanishes. Show that this generalization picks up a new caustic in the cases considered in this chapter (point mass and isothermal sphere). The critical curve is at the Einstein radius and the caustic therefore is at the origin.

3.11 Determine $\kappa(\theta)$ for the isothermal sphere for the case of $\theta_c = \theta_0/4 = 0.1'$. At what value of θ does κ rise above unity? How does this compare to the critical angle in Eq. (3.40)?

3.12 Compute the time delay between two images lensed by a point mass as the source position β varies. Plot this variation as a function of β. Fix $\theta_E = 1.4 \times 10^{-5}$ and the prefactor $(1 + z_L)D_L D_S/D_{SL} = 1$ Gpc.

4

Magnification

Gravitational lensing can change the observed flux received from a background object. To understand how, let us reconsider surface brightness, described in Chapter 1 and Exercise (1.3), the luminosity of an object per angular size, say square arcsecond. A nice property of surface brightness is that it remains constant no matter how far a given object is from us. While the flux (energy per time per area) goes down as the object gets far away, the physical size subtended by a square arcsecond gets larger, so we are seeing photons from a larger physical region, and the two effects exactly cancel (each scaling as the distance from us squared). The energy per time received from a given source then is equal to its surface brightness multiplied by its angular size. Gravitational lensing does not alter the surface brightness of an object but it can change its effective area, so we can receive more light per time from a given object. This is not surprising as it is at the heart of optical lenses: they focus light on a particular point by deflecting more light rays so that they converge on that point. Similarly, mass deflects light as it traverses the universe and can concentrate more or fewer light rays on an observer than would otherwise be seen.

Fig. 4.1 shows a cartoon of an object whose image is enlarged by lensing. In this case, where the source is circular and the circle containing the light gets bigger, the magnification is simply the ratio of the squares of the radii of the unlensed and lensed images. This is a simple example where the source is mapped on to an image via a constant scaling: $\vec{\beta} \rightarrow \lambda\vec{\beta}$. The area of the image is therefore larger by λ^2. More generally, the magnification will be the Jacobian in going from source coordinates $\vec{\beta}$ to the 2D image $\vec{\theta}$.

Magnification is a natural next step after our discussion of multiple images in the last chapter. One feature of multiple images is that they typically lead to us detecting more flux from all the images than if the source were unlensed. This means that, even if we cannot resolve the multiple images directly, we might still learn about the lens by observing the magnification. For example, a very distant galaxy that cannot be resolved, and would be too faint to be observed in the absence of lensing, might be magnified enough to be detected in a survey. This is currently

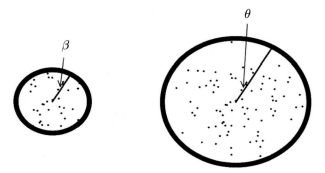

Figure 4.1 Unlensed (left) and magnified (right) object. The surface density (depicted by points) is the same for both objects, but the area from which we receive photons is larger for the lensed object, so it is magnified. In this simple case of a circle, the magnification is just the square of the ratio of the radii of the circles in the image and source planes $(\theta/\beta)^2$.

the best way of finding the most distant and faintest galaxies. Even nearby stars that are easily detectable always appear as point sources with no spatial extent, but they can be magnified in observable ways. We will see several examples where a foreground object briefly intercepts the line of sight between us and a background star, thereby magnifying the star. We do not see the star getting bigger (it remains a point source), but we do observe a change in its brightness and that change can be very informative about the nature of the lens.

4.1 Magnification in Terms of the Distortion Tensor

The flux obtained from a given region is equal to the surface brightness from that region, S, multiplied by the area of the region. Conservation of surface brightness translates to the statement that $S^{\text{un-lensed}}(\vec{\beta}) = S^{\text{lensed}}(\vec{\theta})$. The ratio of the lensed to the unlensed flux, the magnification μ, then is equal to the ratio of the areas from which the photons are observed:

$$\mu \equiv \frac{S^{\text{lensed}}(\vec{\theta})d^2\theta}{S^{\text{un-lensed}}(\vec{\beta})d^2\beta}$$

$$= \left|\frac{d^2\theta}{d^2\beta}\right|. \tag{4.1}$$

Here the ratio is the Jacobian of the transformation from the source area $d^2\beta$ to the image area.

Armed with the lens equation, $\vec{\beta} = \vec{\theta} - \vec{\alpha}$, we can cast the Jacobian in terms of elements of a 2×2 matrix. Explicitly

$$\frac{\partial \beta_i}{\partial \theta_j} = \begin{pmatrix} 1 - \dfrac{\partial \alpha_x}{\partial \theta_x} & -\dfrac{\partial \alpha_x}{\partial \theta_y} \\[2ex] -\dfrac{\partial \alpha_y}{\partial \theta_x} & 1 - \dfrac{\partial \alpha_y}{\partial \theta_y} \end{pmatrix}. \tag{4.2}$$

It appears that there are four derivatives that comprise the matrix, and that the determinant therefore will depend on these four derivatives. However, recall that the deflection angle itself can be written as the derivative of a projected gravitational potential as in Eq. (2.54). Therefore, the two off-diagonal elements of this matrix are equal to one another, and there are only three independent elements of the matrix that describes the effects of lensing. The trace of the matrix is related to the *convergence* κ, while the other two elements are two components of *shear*, γ_1 and γ_2. Explicitly, first the deviation from unity is isolated by defining the *distortion tensor* Ψ_{ij}:

$$\frac{\partial \beta_i}{\partial \theta_j} \equiv \begin{pmatrix} 1 & 0 \\ 0 & 1 \end{pmatrix} - \Psi_{ij} \tag{4.3}$$

and then the elements of the distortion tensor are defined as

$$\Psi_{ij} \equiv \begin{pmatrix} \kappa + \gamma_1 & \gamma_2 \\ \gamma_2 & \kappa - \gamma_1 \end{pmatrix}. \tag{4.4}$$

This immediately allows us to read off the elements of the matrix in terms of derivatives of the projected potential :

$$\kappa \equiv \frac{1}{2c^2} \left(\frac{\partial^2 \Phi}{\partial \theta_x^2} + \frac{\partial^2 \Phi}{\partial \theta_y^2} \right)$$

$$\gamma_1 \equiv \frac{1}{2c^2} \left(\frac{\partial^2 \Phi}{\partial \theta_x^2} - \frac{\partial^2 \Phi}{\partial \theta_y^2} \right)$$

$$\gamma_2 \equiv \frac{1}{c^2} \frac{\partial^2 \Phi}{\partial \theta_x \partial \theta_y}. \tag{4.5}$$

To understand how κ affects an image, consider a line of sight along which $\gamma_1 = \gamma_2 = 0$ and κ is a small positive constant. This corresponds to light from the galaxy passing through an over-dense region. Setting only κ to be nonzero in the distortion tensor leads to the lens equation

$$\frac{\partial \beta_i}{\partial \theta_j} = \delta_{ij} [1 - \kappa]. \tag{4.6}$$

$\kappa > 0$ $\kappa < 0$

Figure 4.2 The effect of nonzero κ on an initially circular background object (the same sized dashed circle in each panel). The left image is when $\kappa > 0$, so the image (solid circle) is dilated, while the right image is smaller because $\kappa < 0$. Both cases assume zero shear.

The lens equation in this diagonal case has only two components:

$$\frac{d\beta_1}{d\theta_1} = 1 - \kappa$$

$$\frac{d\beta_2}{d\theta_2} = 1 - \kappa \tag{4.7}$$

with solution

$$\vec{\beta} = (1 - \kappa)\vec{\theta}. \tag{4.8}$$

If we assume that κ is small, after dividing by $1 - \kappa$, we can Taylor expand so that the image is given by

$$\vec{\theta} = (1 + \kappa)\vec{\beta}. \tag{4.9}$$

So the light that emerged from the edge of the galaxy at source position β_0 is now pushed out and observed at the larger radius of $(1 + \kappa)\beta_0$. That is, the image is dilated, made larger by the convergence κ. Convergence then encodes our argument that the angular size of an object can be enlarged due to lensing. We will therefore see more light from a given source in the presence of dilation: it will be magnified. This case and the case where $\kappa < 0$ are depicted in Fig. 4.2.

We can now calculate the Jacobian of the distortion tensor, and therefore the magnification. The Jacobian of the transformation from $\vec{\beta}$ to $\vec{\theta}$ is the determinant of the matrix on the right-hand side of Eq. (4.3); in terms of its elements, we see that the magnification is

$$\mu = \left[(1 - \kappa)^2 - \gamma^2\right]^{-1} \tag{4.10}$$

where $\gamma^2 \equiv \gamma_1^2 + \gamma_2^2$. Before we work through some examples, note that when the denominator vanishes, the magnification is formally infinite. In reality, of course, the observed magnification is finite; the infinities are removed by the finite extent of the source. Nonetheless, it is true that magnification peaks when the denominator goes to zero. This effect is reminiscent of our discovery in the previous chapter that places where $d\beta/d\theta$ vanishes are special. Indeed, we have arrived at the definition of these special places in the source plane, the caustics. They are defined as places in the source plane where the magnification is formally infinite.

4.2 Point Mass

The simplest nontrivial example of magnification is from a point source mass. We have already computed the potential for this case in Eq. (2.67): $\Phi = \theta_E^2 \ln \theta$ where the prefactor $\theta_E^2 = 4MGD_{SL}/D_S D_L$. Away from the origin, the convergence κ vanishes everywhere so does not contribute to the magnification, so the magnification depends only on the two components of shear. They are obtained by differentiating the potential twice. We find

$$\gamma_1 = -\frac{\theta_E^2}{\theta^4} \left[\theta_x^2 - \theta_y^2\right]$$

$$\gamma_2 = -\frac{2\theta_E^2 \theta_x \theta_y}{\theta^4} \tag{4.11}$$

so the total shear $\gamma = \theta_E^2/\theta^2$. The magnification due to a point mass lens therefore is equal to

$$\mu = \frac{1}{1 - \frac{\theta_E^4}{\theta^4}}. \tag{4.12}$$

We immediately see that when $\beta = 0$, the source perfectly aligned with the lens, and an Einstein ring is formed at $\theta = \theta_E$, the magnification is formally infinite. Practically neither source nor lens is ever a point mass, so this infinity is fictional. To see how this works, consider a source of small but finite angular radius σ directly aligned behind a point mass lens with Einstein radius $\theta_E \gg \sigma$. We can estimate the magnification by appealing to Eq. (4.1). The denominator is the area in the unlensed case, $\pi\sigma^2$. The numerator will be an annulus (instead of an infinitely small ring) of radius θ_E and very small width σ, so the area of the image is $d^2\theta \simeq 2\pi\theta_E\sigma$. Therefore,

$$\mu \simeq \frac{2\theta_E}{\sigma} \gg 1. \tag{4.13}$$

So the magnification is quite real and large for a small source (see also Exercise (4.1)), but certainly finite.

In this case, where the source and lens are perfectly aligned, the caustic is a point at the origin, while the *critical curve*, the curve onto which the caustic is mapped, lies on the Einstein ring at $\theta = \theta_E$. Away from the caustic, $\vec{\beta} \neq 0$ and the magnification is finite. Recall that we found two images produced at $\theta_\pm$, as defined in Eq. (3.7). We can immediately plug these two solutions into Eq. (4.12) to find the magnification as a function of the offset of the source from the lens, β. Fig. 4.3

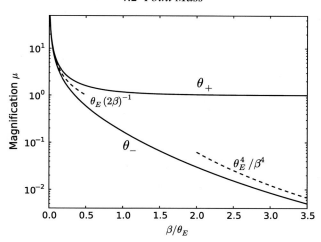

Figure 4.3 Magnification of the two images created by the lensing due to a point mass as a function of the source distance from the lens β in units of the Einstein angle. For large β, the second image – located near the origin – is very demagnified, so is not observable. The dashed curves show the asymptotic expressions for magnification in the large and small β limits.

shows these two magnifications as a function of β in units of the Einstein radius of the lens. It is also illuminating to consider the two limits of small and large β. In the former case, the two images are located at $\theta_\pm \simeq \beta/2 \pm \theta_E$, so they are very close to the Einstein radius on opposite sides of the lens. This leads to magnifications of

$$\mu_\pm \simeq \frac{(\theta_E \pm \beta/2)^4}{(\theta_E \pm \beta/2)^4 - \theta_E^4}$$

$$\simeq \pm \frac{\theta_E}{2\beta} \qquad (\beta \ll \theta_E), \tag{4.14}$$

shown as the dashed curve on the left in Fig. 4.3. We will come to the meaning of a negative magnification in §4.4; for now let's consider the magnitude of μ only. Then, at small β both images are magnified by approximately the same amount. In the large β limit, there is a big difference between the magnification of θ_+, located roughly at $\theta = \beta$ and that of the image at θ_- located close to the origin at $\theta_- \simeq -\theta_E^2/\beta$. The normal image has magnification $\mu_+ \simeq 1 + \theta_E^4/\beta^4$, while the image near the origin has magnification $\mu_- \simeq \theta_E^4/\beta^4$, i.e., very small, again as shown in the figure. Therefore, the second image, already difficult to observe because it must be disentangled from the lens, is highly de-magnified when β is large: we end up seeing a single image at θ_+ very close to the true position with intensity very close to the true intensity.

An interesting and common application is when the two images are too close to be resolved. In that case, the total amplification is the sum of the absolute magnitudes of each separately, so

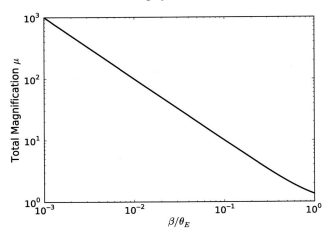

Figure 4.4 Total magnification of the two images generated by a point mass lens as a function of the distance β of the source from the lens in units of the Einstein radius of the lens. When $\beta = 0$, the magnification is formally infinite but is smoothed out by the sizes of the lens and the source.

$$\mu(\beta) = \frac{\theta_+^4(\beta)}{|\theta_+^4(\beta) - \theta_E^4|} + \frac{\theta_-^4(\beta)}{|\theta_-^4(\beta) - \theta_E^4|}. \tag{4.15}$$

In Exercise (4.2), you will show that this is equal to

$$\mu(\beta) = \frac{\beta^2 + 2\theta_E^2}{\beta\sqrt{\beta^2 + 4\theta_E^2}}. \tag{4.16}$$

This total magnification, plotted in Fig. 4.4, depends on the distance between source and lens β; it is apparent that even a lens that does not produce observable multiple images can produce dramatic effects.

4.3 Isothermal Sphere

To compute the magnification induced by a isothermal sphere, we begin with the expression for the deflection angle

$$\vec{\alpha} = \frac{\vec{\theta}\theta_0}{\theta^2} \left[\sqrt{\theta^2 + \theta_c^2} - \theta_c \right]. \tag{4.17}$$

The magnification depends on derivatives of this deflection angle, the convergence κ and the two components of shear, $\gamma_{1,2}$. Comparing Eq. (4.2) and Eq. (4.5), we see that the shear and the convergence are first derivatives of the deflection angle. That is, they are linear combinations of the derivatives:

$$\frac{\partial \alpha_i}{\partial \theta_j} = \delta_{ij} \frac{\theta_0}{\theta^2} \left[\sqrt{\theta^2 + \theta_c^2} - \theta_c \right]$$

$$+ \theta_i \theta_j \frac{\theta_0}{\theta^4 \sqrt{\theta^2 + \theta_c^2}} \left[2\theta_c \sqrt{\theta^2 + \theta_c^2} - 2\theta_c^2 - \theta^2 \right]. \tag{4.18}$$

Let's work through the derivation of κ explicitly and then leave the shear for an exercise. Summing the two diagonal elements gives

$$2\kappa = \frac{\partial \alpha_x}{\partial \theta_x} + \frac{\partial \alpha_y}{\partial \theta_y}$$

$$= \frac{2\theta_0 \sqrt{\theta^2 + \theta_c^2}}{\theta^2} - \frac{2\theta_0 \theta_c^2}{\theta^2 \sqrt{\theta^2 + \theta_c^2}} - \frac{\theta_0}{\sqrt{\theta^2 + \theta_c^2}}. \tag{4.19}$$

Collecting all terms with the square root in the denominator leads to a simple expression for the convergence

$$\kappa = \frac{\theta_0}{2\sqrt{\theta^2 + \theta_c^2}}. \tag{4.20}$$

The shear is computed in Exercise (4.4). We can collect all this to write down the full expression for the magnification of images lensed by the isothermal sphere:

$$\mu = \left[\left(1 - \frac{\theta_0}{2\sqrt{\theta^2 + \theta_c^2}} \right)^2 - \frac{\theta_0^2 \left(2\theta_c \sqrt{\theta^2 + \theta_c^2} - 2\theta_c^2 - \theta^2 \right)^2}{4\theta^4 (\theta^2 + \theta_c^2)} \right]^{-1}. \tag{4.21}$$

4.3.1 Singular Isothermal Sphere

A simple limit of Eq. (4.21) is the singular isothermal sphere, where there is no core so $\theta_c = 0$. In that case, the convergence and shear reduce to

$$\kappa \to \frac{\theta_0}{2|\theta|}$$

$$\gamma^2 \to \frac{\theta_0^2}{4\theta^2}. \tag{4.22}$$

In this special case, the shear and the convergence are equal to one another. Therefore, the magnification is simply equal to

$$\mu = \frac{1}{1 - \frac{\theta_0}{|\theta|}}. \tag{4.23}$$

So, when the image is within the effective Einstein radius $|\theta| < \theta_0$, the magnification is negative, which we will see implies that the image is inverted. Recall that, when $\beta < \theta_0$, two images appear, one at $\theta_+ = \beta + \theta_0$, and the other at $\theta_- = \beta - \theta_0$. The magnification of the two images is

$$\mu(\theta_+) = \frac{\theta_0 + \beta}{\beta}$$

$$\mu(\theta_-) = -\frac{\theta_0 - \beta}{\beta}. \qquad (4.24)$$

The "outer" image then always has positive magnification. When β is large, this is the only image and it is magnified by an amount equal to $1 + \theta_0/\beta$. As β reaches θ_0, a second image appears, initially with $\mu = 0$, meaning it is not detectable, but then growing as β moves towards the origin and eventually growing arbitrarily large. This image is always inverted.

The caustic, the place where the magnification blows up, is only at $\beta = 0$. The image of that source is of course an Einstein ring at θ_0; this is called the critical curve. Note that $\beta = \theta_0$ is a special place in the sense that a new image appears, but it is *not* a caustic because the determinant, and therefore the magnification, is finite there. This, however, is an artifact of the fact that the profile goes to infinity as $r \to 0$; when the cored isothermal sphere is considered next, we will see that generally crossing a caustic leads to the introduction of two new images.

4.3.2 Cored Isothermal Sphere

Recall that a cored isothermal sphere produces a single image as long as the source position β is larger than some critical value β_{cr}. When $\beta < \beta_{cr}$, three images appear. Let's consider an example, culled from Fig. 3.7, where the core radius θ_c was set equal to $\theta_0/4$. For concreteness, we first examine the three images produced when $\beta = 0.1\theta_0$. As depicted in the figure, the positions of these images (shown as circles) are $\theta_{1,2,3}/\theta_0 = -0.54, -0.11, +0.85$. Fig. 4.5 shows the positions and magnifications at these three points.

The magnifications of the two new images – to the left of the lens in Fig. 4.5 – are -10.75 and 1.5, while the original image is magnified by a factor of 11.3. The observables are the positions of the images and the flux ratios. In this case, the

Figure 4.5 Magnification of three images produced by a lensing isothermal sphere with $\theta_c = \theta_0/4$. The position of the lens is shown with an "x"; the source with a square; and the lensed images with circles. The sizes of the circles depict how magnified they are. Closed circles have positive magnification, while the open circle has negative magnification, as explained in §4.4.

flux ratio of the two images to the left of the lens is over 7, while the flux ratio of the brightest images is about 1.05. The negative sign of μ for the leftmost image, encoded by the open circle, will be discussed in §4.4.

More generally, the flux ratios can be plotted as a function of the source position. The two images that appear only when β is within the caustic always have a large flux ratio. The image closer to the center is fainter than the other two images. For two reasons then – it is very faint and it overlaps with the lens – it is rarely observed. This leads to the rather odd state of affairs that gravitational lensing contains a theorem that states the number of images must be odd, but the observed number of images in most discovered systems is even.

Fig. 4.6 shows the magnification as a function of source–lens distance for a particular value of θ_c. Moving from right to left in the figure, when the source is far from the lens, the single image is slightly magnified, but as β decreases, the magnification increases. When β hits its caustic (in this case $0.17\theta_0$), two new images with formally infinite magnification appear. As β moves within the caustic, the magnifications of the two new images diverge, with the outer image becoming brighter and the inner image eventually not magnified at all. Moving towards the caustic at $\beta = 0$, the original image and the new outer image are infinitely magnified and they form the Einstein ring.

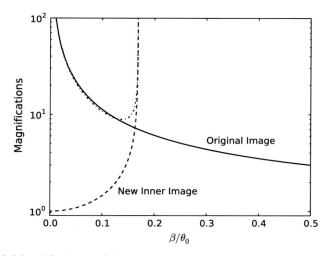

Figure 4.6 Magnifications of the images produced by an isothermal sphere with $\theta_c = 0.25\theta_0$ as a function of the distance of the source from the lens (β). When the source is far from the lens, there is only a single image, but it is magnified. As the source approaches the caustic – in this case, $\beta = 0.17\theta_0$ – two new images appear with infinite magnification. As the source gets closer to the center, one of the new images becomes much brighter than the other, as does the original image. By the time $\beta \to 0$, the two brightest images form an Einstein ring.

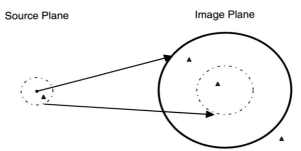

Figure 4.7 Caustics in the source plane mapped to critical curves in the image plane for the isothermal sphere. The triangles in both planes refer to the source and image positions corresponding to Fig. 4.5.

The caustics are the two places where the magnification is infinite: at $\beta = 0$ and – in the case shown in Fig. 4.6 – at $\beta = 0.17\theta_0$. Each of those caustics is mapped onto a curve in the image plane. Fig. 4.7 depicts the caustics on the left: the point at the origin is mapped to the Einstein ring in the image plane, while the inner caustic, the crossing of which leads to multiple images, is mapped onto a critical curve with smaller radius in the image plane. Also shown in both planes are one source position ($\beta = 0.1\theta_0$, as was drawn in Fig. 4.5) and the positions of the three images. Note that the two new images on the far side of the lens are on opposite sides of the critical curve. This is generally the case: two new images form on the critical curve when the source resides on a caustic; as the source moves inwards, the two new images separate, moving to either side of the critical curve, but as long as they are relatively close to one another, they have similar magnifications.

4.4 Parity

We have noticed several times that the magnification of an object, as expressed for example in Eq. (4.10), can apparently be either positive or negative. What is the meaning of a negative magnification? To answer this, let's consider the question in the context of the isothermal sphere, and to be concrete let's continue to focus on the example depicted in Fig. 4.5. There, three images are shown, one of which has negative magnification. The sizes of the circles in Fig. 4.5 simply express the amplitude of the magnification, not any physical extent. So instead let's redraw two of the images there showing their physical extent. This will prove useful to understand the sign of the magnification.

Consider then the two images of the one-armed alien in Fig. 4.8. The image to the right of the source is the one we have encountered several times already: even if only one image is produced, it is typically moved in the image plane farther away

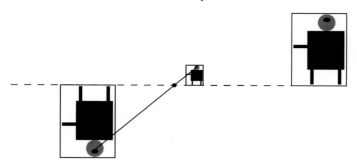

Figure 4.8 Images of a one-armed alien by an isothermal lens. The (small) source is just to the right of the lens. Two of the three images are shown. The far-right image is magnified with the alien right-side up. The leftmost image has negative magnification so appears inverted; it too is much larger than the original. The inverted nature of this image means that it has *negative parity*. The solid line shows the axis along which the eye of the source and image are aligned; as we have seen, this line intersects the lens. Note that due to the flip of the image about the horizontal axis, the alien appears left-handed.

from the lens. Every point in this extended image moves farther from the lens, so the apparent image looks very much like the unlensed source, just bigger.

Consider now the leftmost image in Fig. 4.8. It is straightforward to understand why the alien is upside-down: since the lens is assumed to be spherically symmetric, the alien's feet remain on the line connecting them to the lens, the horizontal dashed line in the figure. For the same reason, the light rays emanating from the alien's head are also imaged on a line that connects the head to the lens. Therefore, as indicated by the solid line, the head appears below the line connecting the lens to the feet. So the alien appears upside down. The placement of the hand is a little trickier. The unlensed source picture shows that the alien is right-handed. Therefore, in this case, its hand is closer to the lens than is the rest of its body. But we know that, for an isothermal sphere, a source closer to the lens leads to this image farther away from the lens (recall Fig. 3.7). Therefore the alien's hand in this image appears farther away from the lens than the rest of its body: it appears left-handed. The flipping of the image is captured by the fact that the magnification is negative. Indeed this is the meaning of a negative magnification: the image is inverted. This image is said to have *negative parity*.

There is a slightly more sophisticated way to think about the magnification and the distortion tensor. The leftmost image in Fig. 4.8 is flipped around the x-axis. If we think of $\delta_{ij} - \psi_{ij}$ as a 2×2 transformation matrix, this corresponds to one of its eigenvalues being negative (at the position of the image) and one positive. Therefore its determinant is negative. Consider the third image now, as depicted in Fig. 4.9: you should be able to walk through the same qualitative arguments to show

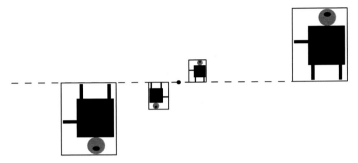

Figure 4.9 Same as Fig. 4.8, but this time including the third image. This image has total positive parity, but has been flipped about both the horizontal and vertical axes, so has two negative eigenvalues. The alien therefore appears right-handed in this image.

that this image is as it appears: flipped about both the horizontal and vertical axes. The double-flip means that the alien now appears right-handed again. It also means that the transformation matrix has two negative eigenvalues, so its determinant is positive. The parity of this image therefore is positive.

4.5 Detection of Magnification

The observation of multiple images is a striking manifestation of gravitational lensing. Magnification, on the other hand, is not necessarily observable at all, because the observer typically does not know how bright the object would be in the absence of lensing. If we see a very bright object, it may be intrinsically bright or close to us or it may be intrinsically faint but magnified by an intervening lens. Indeed, a theme running through this book is: how can we detect the effects of lensing with limited prior knowledge of the properties of the sources?

For magnification, one way to detect the effect is to compare the brightnesses of galaxies behind massive lenses to those in random positions. The idea then is to compare the brightnesses of background galaxies behind known lenses with the brightnesses of background galaxies with similar properties not aligned with a foreground lens. The amount of magnification will then tell us something about the masses of the lenses.

Let's consider a concrete example using numbers similar to those obtained in a detection of this effect. Consider lenses described by a singular isothermal sphere profile with velocity dispersion $\sigma = 350$ km/sec, a distance of 1 Gpc away from us. Also, consider source galaxies all a distance of 1.7 Gpc from us. From Eq. (3.28), the Einstein radius in this case is equal to $\theta_0 = 1.5''$. Even telescopes that can probe very deeply, such as the Hubble Space Telescope, typically see fewer than 50 galaxies per square arcminute. Since the area within the Einstein radius in this case is 0.2 percent of a square arcminute, the number of background galaxies within the Einstein radius expected is much less than one. So to detect the effect of magnification in this case, one must observe galaxies outside the Einstein radius: those that are magnified only slightly. In that case, we are in the regime of weak lensing, and the magnification of the lone image becomes

$$\mu \simeq 1 + 2\kappa$$
$$= 1 + \frac{\theta_0}{|\theta|}. \tag{4.25}$$

To translate this into an observable, we introduce the *apparent magnitude m*, which quantifies the flux observed from an object. Apparent magnitude is related to absolute magnitude and distance, but more directly to the flux from an object, with

$$m \equiv -2.5 \log_{10} \left(\frac{F}{F_{\text{fiducial}}} \right); \tag{4.26}$$

here F_{fiducial} is some fiducial value that will turn out to be irrelevant. Since the flux from a lensed object F is larger than F_0, its unlensed value, the apparent magnitude m will be smaller than the apparent magnitude if unlensed, m_0:

$$m - m_0 = -2.5 \log_{10} \left(\frac{F}{F_0} \right)$$
$$= -2.5 \log_{10}(\mu). \tag{4.27}$$

In the weak lensing regime, we can Taylor expand the log to obtain

$$m - m_0 \simeq -\frac{2.5}{\ln(10)} \frac{\theta_0}{\theta}. \tag{4.28}$$

In a given annulus with radius θ and width $\Delta\theta$, then the *signal* in our example – i.e. the difference between the magnitudes of the lensed and unlensed galaxies – is $1.6''/\theta$. A typical value of θ for which we can obtain a fair number of background galaxies is of order an arcminute, so the magnitude difference that serves as the signal is much less than one. The *noise* is the intrinsic scatter in the magnitudes of the galaxies; i.e., not all galaxies have apparent magnitudes exactly equal to m_0. If the spread is $\sigma_m = 1$, then the situation appears hopeless, as the signal-to-noise is much less than one. Statistics comes to the rescue, though: if many background galaxies can be measured in the annulus, the noise goes down as the square root of this number (see, e.g., Exercise (4.8)). Assuming then $n = 50$ galaxies per square arcminute, the total signal to-noise in an annulus then is

$$\left. \frac{S}{N} \right|_\theta = \frac{1.6''}{\theta} \sqrt{n 2\pi \theta \Delta\theta}$$
$$= 0.5 \left[\frac{\Delta\theta}{\theta} \right]^{1/2}. \tag{4.29}$$

Taking angular bins with fractional width $\Delta\theta = 0.25\theta$ shows that we are not quite there: the signal-to-noise for a single lens, even using multiple background galaxies, is estimated to be of order 0.25.

Schmidt et al. (2012) detected the effect of magnification by using not one but many lenses and beating down the noise yet further. They selected 61 groups of galaxies found in X-ray data as lenses, together with 250,000 background galaxies from the Hubble Space Telescope over 1.4 square degrees. The multiple lenses beat down the noise by a further factor of $\sqrt{61}$, giving an estimate of the signal-to-noise of 2 for each angular bin.

Fig. 4.10 shows this detection. The filled points denote the measured surface density, which can be related to κ via the critical surface density (Exercise (4.6)). Note that the bins are roughly as described with width $\Delta r / r = \Delta\theta / \theta \simeq 0.25$, and the ratio of the amplitude of the signal to the size of the error bar – the signal-to-noise – in each bin is about 2, in agreement with our estimate. Hidden in this description is that they used not just the magnitudes but also the sizes of the galaxies, and this is important in order to define a sample with a relatively tight spread in magnitudes (recall that we assumed $\sigma_m = 1$). Also shown in the figure are estimates using the shapes of the background galaxies, a technique that we will consider in Chapters 6 and 7. Currently, measurements using shapes are

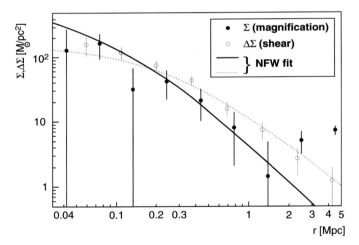

Figure 4.10 Surface density ($\Sigma = \Sigma_{cr}\kappa$) of 61 foreground lenses inferred from magnification (closed circles with best bit solid line) as a function of distance from the lenses (Schmidt et al., 2012). Open circles and the dashed curve show measurements of a slightly different quantity to be defined later, $\Delta\Sigma$ inferred from the shapes of background galaxies. The solid curve shows the best fit, not from an isothermal sphere profile but rather from one proposed by Navarro et al. (1997) (NFW) that has been shown to fit numerical simulations (see Exercise (3.6) for some details).

more precise, but shapes become increasingly difficult to measure as telescopes go deeper and measure fainter (and therefore smaller) objects. It is possible that future weak lensing measurements will rely not just on shapes but on magnification as well.

4.6 Using Lenses to Find Distant Galaxies

By using observables such as the positions of multiple images and their magnitudes, modelers can estimate the mass profiles of a lens (see Exercise (4.7) for an example). Armed with these *mass maps*, one can then identify the critical curves, the positions on which background galaxies will be significantly magnified. This opens up a new avenue of research: hunting on critical curves for very faint background objects, objects so far away that they would not be seen in the absence of the magnification. Effectively, lensing then acts as a second telescope, empowering astronomers to find very distant objects. Indeed, the most distant galaxies in the universe have been found by hunting near critical curves.

Fig. 4.11 shows one example of this from the Hubble Space Telescope. The numbers denote all the foreground objects used to make the mass map, and the red circles show the positions of the three images. The lensed galaxy is very faint and would have apparent magnitude $m = 30$ in the absence of magnification. Even the Hubble Telescope could not have detected objects that faint (its limiting magnitude for its deepest scans is of order 29). The two leftmost images have been magnified by a factor of order 10, though, so the observed apparent magnitudes are between 27 and 28: within reach of the Hubble Telescope.

The positions of the critical curves depend on how far away the source is from the lens. We can see this explicitly in both of the simple cases treated so far, the point mass and the isothermal sphere. In both cases, the Einstein radius scales as $(D_{SL}/D_S)^{1/2}$ so gets larger as the source gets farther away. The equations that determine the critical curves can all be scaled to be functions of θ/θ_E, so as the Einstein radius gets larger, the critical curves move away from the center of the lens. This means that critical curves far from the lens are an indication of very distant sources.

Redshift as a distance indicator

The spectrum of an object such as a galaxy encodes information about how far away that object is. The spectrum typically includes emission and absorption lines stemming from a variety of atomic transitions. Because the universe is expanding, objects far away from us are moving rapidly away from us, and therefore those lines are *redshifted*. For example, the Hα emission line corresponds to radiation emitted during a transition

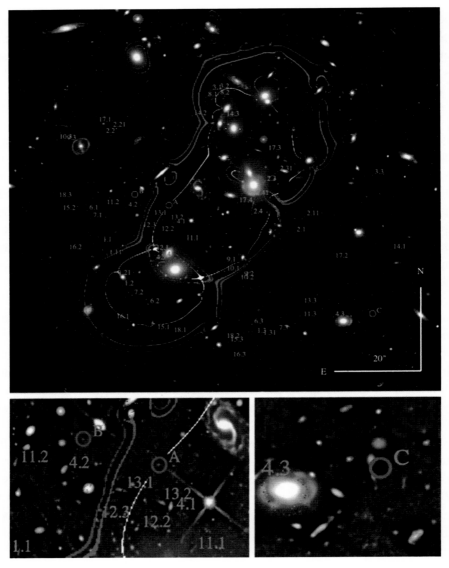

Figure 4.11 A very distant galaxy, detected only because its (three) images (red circles) are magnified by a factor of ten. Top panel shows the full field, with numbers representing galaxies used to construct the mass map; critical curves determined by multiple imaging of closer objects shown in different colors depending on the distance of the source; and the three images of the background galaxy itself indicated by the red circles (Zitrin et al., 2014). Different colored critical curves correspond to different distances to the background galaxy. Bottom panels zoom in to the three images; the bottom-left panel shows the two nearby images, which should be nearly equidistant from the critical curves. This rules out the blue and white curves that correspond to background redshifts of 1.3 and 2. (See color plates section.)

of the electron in atomic hydrogen from the $n = 3$ to the $n = 2$ state. In the lab, this line has a wavelength of 656 nm. If a galaxy is far away, and therefore moving rapidly away from us, the line will show up at a longer wavelength. For example, if it is at wavelength $\lambda = 1000$nm, then the galaxy's redshift is determined by $1+z = 1000/656$ so $z = 0.52$.

Astronomers often take this information from the spectrum and refer to a "galaxy at redshift z." Because objects farther away emit light that has been redshifted more, this statement captures information about the distance of the galaxy from us. However, the galaxy presumably existed for some time before and after the light ray we detect was emitted (rays emitted earlier have passed us by and those emitted later have not yet reached us). So the statement that an object is at a particular redshift is a statement about the object at the time the light we observe was emitted. We will get more quantitative in §7.5, but for now the take away is that the redshift serves as a proxy for the distance between us and the emitting object.

The bottom-left panel of Fig. 4.11 is a zoom-in on two of the (three) images of the faint source galaxy. The colored curves are critical curves that are determined from all the other observations (numbers in the figure are images used to construct the critical curves). These critical curves depend on the redshift of the source (see shaded box), with the white curve corresponding to source redshift $z_s = 1.3$ and blue to $z_s = 2$. Each of those is excluded simply because the two extra images should be roughly equidistant from the critical curve. Even the green curve corresponding to $z_s = 3.6$ seems disfavored, with the red ($z_s = 10$) apparently a better fit. This is an extraordinary way to determine the redshifts (and therefore distances) to source objects. In this case, the constraint that z_s is most likely greater than 3, combined with information from the colors of the faint object, place the source at $z = 10$.

4.7 Multiple Image Systems and Flux Anomalies

Astronomers have discovered many systems in which there are four images of the same source object. Fig. 4.12 shows one example. This system was first discovered in radio observations at the Very Large Array in New Mexico and then observed in different wavelengths on the Keck Telescope and the Hubble Space Telescope. The suite of observations confirms that the blobs labeled A, B, C, and D are indeed four images of the same background object lensed by a foreground object at the point labeled E in the figure.

The image separations are on the order of an arcsecond, as expected from our calculations above about the Einstein radii of lensing galaxies. The fluxes have

Magnification

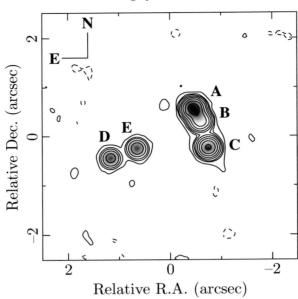

Figure 4.12 Four-image gravitational lens system B2045+265 (Fassnacht et al., 1999) as observed by the Very Large Array at 8.5 GHz. This system was discovered in the radio and then followed up at other wavelengths. The images labeled A, B, C, and D are the four images of the same background compact source. The radio emission at the point labeled E is likely due to the lens galaxy.

been measured accurately and, although there is some variation depending on the wavelength, image A is roughly twice as bright as image B, and the flux from image C lies in-between its two nearest neighbors. This will turn out to be a puzzle.

Fassnacht et al. (1999) introduced a mass model of the lens to try to account for the observed positions and fluxes of the images. The model is shown in Fig. 4.13. The dashed curve shows the critical curve, which is similar to the ones we have been exploring in that it is smooth, but clearly it is not symmetric. Indeed, the model includes a singular isothermal sphere plus an asymmetric term with a potential $\Phi \propto \cos(2[\phi - \phi_0])$ where $\tan \phi \equiv \theta_y/\theta_x$. This second term provides the elliptical shape of the critical curve. The caustic corresponding to this asymmetric potential is shown by the solid curve: apparently, the asymmetry leads to more structure in the caustic with four cusps.

One of the free parameters in the model is the true source position, the best fit of which is shown as the filled triangle in Fig. 4.13. So the model requires the source to be very close to the caustic; if it had been outside the solid curve, there would have been only two visible images, not four, as each crossing of a caustic introduces two new images. (There is a fifth image predicted, at the center, but this is – as usual

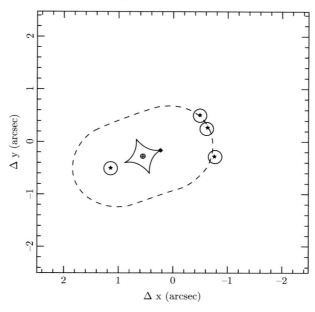

Figure 4.13 Model of the system shown in Fig. 4.12 (Fassnacht et al., 1999), with the location of the center of the lens denoted by the circled plus sign. The caustic curve in the model is depicted by the solid cuspy curve. The dashed ellipse shows the critical curve. The true position of the source in the model is shown by the filled diamond, while the stars show the predicted positions of the four images. The positions of the actual images are slightly uncertain, as depicted by the open circles. Note the good agreement between the predictions of the model and the positions of the images.

– too faint to be seen.) Perhaps not surprisingly, given our investigations into the isothermal sphere, images A and C are close to one another. When the source is right on top of the caustic, the images form at the same location and then move away from one another as the source moves away from the caustic. So, indeed, the positions of the images in the model agree well with those in the data.

The problem with the model, not only in this case but in a number of four-image lenses, is that it predicts that flux from image B should equal the sum of the fluxes from A and C. The observations show that B is much fainter than expected. Indeed, this is a robust prediction of any model with a relatively smooth potential. There have been a number of proposals for improving the models so that they agree with the data; perhaps the most interesting is that the lens galaxy is populated with small clumps of dark matter that destroy the smoothness of the potential. Dalal and Kochanek (2002) proposed that the flux anomalies could be used to *measure* the amount of dark matter clumps in a lensing galaxy, a measurement that could be compared with the predictions from cosmological simulations.

Suggested Reading

The textbooks mentioned in previous chapters do a good job of covering magnification. The first step in moving beyond the isothermal sphere used extensively here as an example is to work through similar steps for the NFW profile (Navarro et al., 1997); a clear paper that does this is Wright and Brainerd (1999). Then, to move beyond lenses with spherical symmetry, pick an early paper by Roger Blandford; again, the paper by Blandford and Narayan (1986) is perhaps the best place to start.

The possibility of detecting the effect of magnification by measuring brightnesses in the weak lensing regime, as outlined in §4.5, has been revitalized with the detections of, e.g., Huff and Graves (2014) and Schmidt et al. (2012). A nice, short review of some of the subtleties and detections is contained in section 3.3 of Rhodes et al. (2015).

The detection of distant, faint galaxies using foreground lenses to magnify the images dates back as far as Franx et al. (1997) and is described beautifully in the paper by Zitrin et al. (2014), treated in §4.6.

Exercises

4.1 Redo the calculation in the text surrounding Eq. (4.13) more carefully by defining the magnification of the entire source as

$$\bar{\mu} \equiv \frac{\int d^2\beta \, \mu(\beta)\mathcal{S}(\beta)}{\int d^2\beta \, \mathcal{S}(\beta)}, \qquad (4.30)$$

assuming a surface brightness that is Gaussian with width σ.

4.2 Starting from Eq. (4.15), use the expressions for $\theta_\pm$ for a point mass to derive Eq. (4.16).

4.3 Show that, far from a point mass lens, the magnification of an image is

$$\mu_+ \simeq 1 + \frac{\theta_E^4}{\beta^4} \simeq 1 + \frac{\theta_E^4}{\theta^4}. \qquad (4.31)$$

4.4 Show that the total shear squared for an isothermal sphere is equal to

$$\gamma^2 = \frac{\theta_0^2}{4\theta^4(\theta^2 + \theta_c^2)} \left[2\theta_c\sqrt{\theta^2 + \theta_c^2} - 2\theta_c^2 - \theta^2 \right]^2. \qquad (4.32)$$

4.5 Create a plot like Fig. 4.6 for the value $\theta_c = 0.05\theta_0$.

4.6 Calculate $\Sigma(r)$ for the isothermal profile described in §4.5, using the form of κ in Eq. (4.20). For this, you will need to determine Σ_{cr} from the given distances and velocity dispersion and convert angle θ to distance r using the distance to the lens of 1 Gpc.

4.7 One of the key goals in strong lensing studies is to make mass maps. These lead to the identification of critical curves that are essential for applications like the one described in §4.6. Use the following data to create a mass map. The data set consists of three images with positions relative to the center of the lens and magnitudes equal to:
1. $\theta = +1.767''$ and $m = 18.06$
2. $\theta = -0.082''$ and $m = 22.6$
3. $\theta = -0.989''$ and $m = 18.47$
Fit this data set with a cored isothermal sphere, determining the two free parameters. Where is the critical curve?

4.8 Suppose one wanted to estimate the percentage of the vote a candidate is expected to get in the Iowa caucuses. Plot the χ^2 from (i) a single poll with an estimate of 45 percent for the candidate and from (ii) ten polls with estimates of $p_i = [44, 42, 43, 46, 43, 49, 48, 43, 43, 43]$ In both cases, assume that each poll has an intrinsic uncertainty of $\Delta p = 3$ percent and assume

$$\chi^2(p) = \sum_{i=1}^{N} \frac{(p - p_i)^2}{(\Delta p)^2}. \tag{4.33}$$

Show that the χ^2 for the two different cases looks like Fig. 4.14.

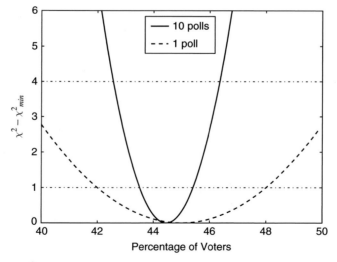

Figure 4.14 χ^2 for the percentage of votes expected by a given candidate, computed from a single poll with a 3 percent margin of error (dashed curve) and from ten polls with the same margin of error. The horizontal lines depict the rough 1- and 2-σ limits when χ^2 exceeds its minimum by 1 and 4, respectively. Note that the curvature of the χ^2 is much steeper when more data are used. This suggests the rule of thumb that the signal-to-noise increases as $\sqrt{N}$.

4.9 Simply measuring the magnitudes of background galaxies is not enough to detect magnification. To understand why, first compute the mean apparent magnitude of a flux-limited galaxy sample drawn from a power law distribution

$$\frac{dn}{dm} = Am^{\alpha}. \tag{4.34}$$

Flux-limited means that only galaxies brighter than m_{limit} are in the sample. Now redo the calculation assuming that all galaxies are magnified by an amount $\mu > 1$. Show that the mean magnitude remains the same! So a flux-limited power law distribution does not work. Typically, what needs to be done is to define a galaxy sample that has fairly uniform properties, so does *not* come from a power law distribution. Hint: integrate over the *un-lensed m* with an upper limit of $m_{limit} + 2.5 \log_{10} \mu$ since all galaxies with un-lensed m less than this will now be included in the sample.

5

Microlensing

Until now we have been thinking of the source and lens as being located at fixed positions, so that the effects – e.g., multiple images and magnifications – will be fixed in time. There are some situations, though, where this is not true, where the relative positions of the lenses and sources change with time. In those cases, the magnifications, for example, will also change with time. This is potentially observable: if a source is detected with a given flux at one time and a different flux at another, the differences might be caused by the lens moving and creating more magnification at different times.

The name for this phenomenon is *microlensing* because it was first coined in observations of distant QSOs whose light was lensed by an intervening galaxy. If the light rays passed close enough to a single star in the lensing galaxy, then the image was perturbed in a time-dependent fashion that will occupy us in this chapter. We can estimate the deflection angle for this type of event: $4MG/D_L c^2$ for a solar mass star a cosmological distance of 1 Gpc from us is 3×10^{-6} arcseconds. The deflection angles are very small; hence the moniker, microlensing. However, the name is often used to describe any time-dependent change in flux even if the deflections are not at the 10^{-6} arcsecond level.

Fig. 5.1 gives a generic example, with the background source traversing through the Einstein radius of the lens. Of course, the motion is relative, so this is equivalent to the more likely case where the lens is moving and the source remains fixed. In either case, the magnification of the source will reach a maximum value when the lens and source are closest to one another, an angular distance β_{min} away. An observer will detect a rise and then fall of the flux from the object.

In this simple example of a point mass lens, the magnification as a function of the relative position of the two objects is shown in Fig. 5.2 for two different values of β_{min}. This characteristic signature of brightening and then dimming offers an excellent prospect for detecting a object moving between us and a background source. Consider as an example a solar mass lens in our galaxy a distance of 5 kpc from us, so that its Einstein radius is of order 10^{-3} arcsecond. If it is moving with

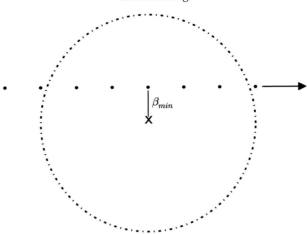

Figure 5.1 Simple example of microlensing. Background source (dot) moves within the Einstein radius (dot-dashed circle) of the lens ("x") and then moves out. In the course of this trajectory, its flux is first magnified and then resets to normal.

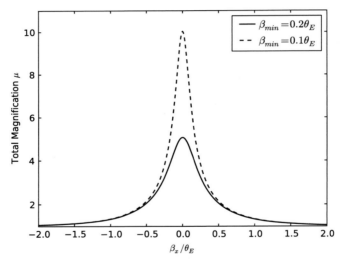

Figure 5.2 Total magnification of a background image as a function of its motion along the β_x-axis. The angular distance of closest approach corresponds to the (fixed) β_y distance, called $\beta_{\min}$ in Fig. 5.1. As the angular distance between the source and lens changes, the brightness of the background source changes with time, with peak magnification depending on the value of $\beta_{\min}$.

a (typical) velocity of 100 km sec^{-1} in the plane perpendicular to the line of sight, then, over the course of a year, it will therefore traverse an angular distance equal to four times its Einstein radius. Therefore, a well-aligned background object will indeed experience this characteristic magnification pattern.

A fundamental principle of lensing is that the way in which light paths are distorted does not depend on the wavelength of the light. Therefore, these magnification curves should look the same at all wavelengths. This feature is very useful, practically, since one needs to distinguish microlensing from objects with intrinsic variability. Often, that variability does depend on wavelength, so multiple wavelength observations are tools to distinguish intrinsic variability from microlensing.

We will focus on two very important questions that can be addressed with microlensing: the identity of the dark matter that permeates the Galaxy (and the rest of the universe) and the possibility that other stars in the Galaxy also have planets orbiting them just as the Sun does.

5.1 Dark Matter in the Galaxy

The vast majority of visible matter in the Milky Way ("the Galaxy") resides in a disk with thickness of order 1 kpc and radius of order 5 kpc. Fig. 5.3 offers

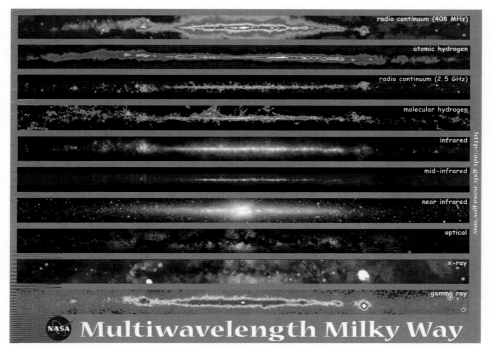

Figure 5.3 View towards the Galactic center of the Milky Way at different wavelengths. The near-infrared radiation comes predominantly from giant stars and shows a bulge that tapers off at ∼ 1 kpc. Emission from hydrogen, both molecular and atomic, traces the gas and extends further, out to about 5 kpc. There is a clear fall-off at larger radii, indicating that most of the emitters are confined to a disk with a radius of about 5 kpc. (See color plates section.)

excellent confirmation of this idea, which goes back several centuries. Each row is a view towards the Galactic center at a different wavelength. The optical wavelength shows very little structure because there is so much dust along the line of sight that most of the starlight is absorbed or scattered. Longer wavelength light, on the other hand, scatters less off the relatively small dust particles (typically less a micron in size), so the near-infrared view offers a much cleaner picture of the distribution of stars. We see a bulge with a radius of order 1 kpc and then a sharp drop in the stellar density. The gas is traced by emission from neutral hydrogen, both atomic and molecular, and both of those views indicate that the gas is confined to a disk and its density falls off rapidly beyond about 5 kpc.

The expectation is that this distribution of matter would produce a gravitational field that falls off as $1/r$ as r gets larger than 5 kpc. Admittedly, the field produced by a disk is complicated, but far from the edge of the disk, an excellent approximation is that the field is equivalent to that produced by a single point at the center with mass equal to the total mass within the disk. Equating the gravitational force (which therefore falls off as r^{-2}) with the centripetal acceleration of a test object (v^2/r), we expect the classic Keplerian result: stars further away than 5 kpc should be rotating around the center of the Galaxy with velocities that fall off as $r^{-1/2}$. Fig. 5.4 shows a compilation (Sofue, 2012) of velocities obtained in a variety of ways. It is clear that there is no fall-off: the rotation curve is flat and disagrees with the Keplerian prediction.

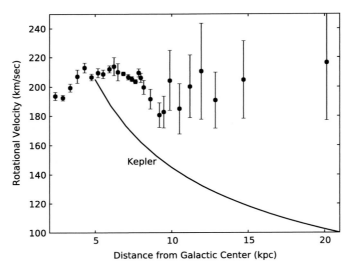

Figure 5.4 Rotation curve of the Milky Way, data accumulated by Sofue (2012). Solid curve shows the expected fall-off as $r^{-1/2}$ starting from the place where the visible matter drops, at around $r = 5$ kpc.

The best explanation for this anomaly is that there is dark matter in the Galaxy, distributed in a *halo* (spherical distribution) that extends far beyond the visible region but is unseen. To obtain a flat rotation curve, the mass enclosed within a radius r, $M(r)$, must satisfy:

$$\frac{MG(r)}{r^2} = \frac{v^2}{r}; \tag{5.1}$$

i.e., the mass must increase linearly with r. So, to the extent that the Galaxy has a flat rotation curve, the dark matter halo must extend out quite far with density falling off roughly as r^{-2} (so that $\int d^3 r \rho(r) \propto r$). We are embedded in the Milky Way, so it is difficult to measure our rotation curve, but there are data points that go out as far as hundreds of kpc without any clear indication of appreciable decline, so it is likely that the dark matter halo in the Galaxy extends out very far and – as a corollary – there is about ten times as much mass in the Galaxy in dark matter as there is in gas or stars.

The notion that dark matter permeates the Galaxy would be taken much less seriously if there was not quite a bit of other evidence for it. In terms of rotation curves, it is not just our Galaxy that has a rotation curve that appears inconsistent with its visible matter: virtually all galaxies for which rotation curves can be measured show evidence of stellar velocities far larger than would be expected at their outskirts. So, from rotation curves alone, the dark matter hypothesis should be taken very seriously. And, as we progress through lensing applications, we will encounter more and more evidence for dark matter.

This raises the tantalizing question: what is it? Thousands of articles and dozens of books have been written about this, but the short answer is that the nature of the dark matter is unknown. We will see in the next section that lensing can provide information about one possibility: that the dark matter consists of compact massive objects. Before turning to that, it is worth noting that the most popular idea for dark matter is that it consists of weakly interacting massive particles, so-called WIMPs. These are elementary particles, which mostly rotate around the center of the Galaxy independently, each with a mass within a factor of a hundred or so of the proton. So, if the dark matter consists of WIMPs, there will be relatively few dark compact objects with masses of order that of the Sun.

5.2 MACHO Microlensing

One candidate for dark matter is a Massive Compact Halo Object, or a MACHO. MACHOs could be neutron stars, black holes, or planets that do not emit much light but that do contain substantial mass. Microlensing of background stars would be a signature that the dark mass in the Galaxy consists of relatively large objects.

Imagine a fixed background object, a star say, and a foreground point mass lens moving in our Galaxy. The configuration is determined by the distance of closest approach, β_{min} in Fig. 5.1. As the lens moves across the sky, the angular distance between the true position of the source and the lens, β, changes: it is initially very large, decreases to a minimum value of β_{min}, and then increases again as the lens moves away from the line of sight. We have already determined the magnification of a source lensed by a point mass as a function of β in Eq. (4.16). If we define the variable $u \equiv \beta/\theta_E$, then

$$\mu = \frac{u^2 + 2}{u\sqrt{u^2 + 4}}. \tag{5.2}$$

This is precisely what is plotted in Fig. 5.2.

Eq. (5.2) is deceptively simple: it hides the facts that u changes with time as the MACHO traverses the line of sight and that the Einstein radius depends on how far the MACHO is from us and from the source. Explicitly, the angular distance between the MACHO and the background star is $\beta = |\vec{\beta}_M - \vec{\beta}_*|$. Since the background star is fixed in space, let us choose the star to be at the origin. The time variation of β depends on the minimum distance β_{min} and on the transverse velocity of the lens, $v_\perp$. The total angular distance is β_{min} added in quadrature with the transverse distance from that minimum position. We can then set $t = 0$ as the time at which $\beta = \beta_{min}$ so that the transverse angular distance is $v_\perp t/D_L$. Therefore, the dimensionless variable u that governs the magnification is

$$u = \frac{\sqrt{\beta_{min}^2 + v_\perp^2 t^2/D_L^2}}{\theta_E}. \tag{5.3}$$

We can immediately obtain from Eq. (5.3) an estimate of the duration of the magnification. Roughly, the object will be magnified as long as the numerator, the angular distance between the lens and the source, is smaller than the Einstein radius, or $u < 1$. If β_{min} were equal to zero, then this would hold as long as $v_\perp t/(D_L\theta_E)$ were less than one. This suggests defining the event duration as

$$t_d \equiv \frac{D_L\theta_E}{v_\perp}$$

$$= 87\,\text{days}\,\frac{200\,\text{km/s}}{v_\perp}\,\frac{D_L}{10\,\text{kpc}}\,\frac{\left(\frac{4MGD_{SL}}{D_S D_L c^2}\right)^{1/2}}{10^{-3}\,\text{arcsec}} \tag{5.4}$$

where the second line inserts the definition of the Einstein radius of a point mass. It is cleaner to write the scaling in terms of the mass of, and distance to, the MACHO:

$$t_d = 78\,\text{days}\,\frac{200\,\text{km/s}}{v_\perp}\left(\frac{M}{M_\odot}\right)^{1/2}\left(\frac{D_L}{10\,\text{kpc}}\right)^{1/2}\left(\frac{D_{SL}}{D_S}\right)^{1/2}. \tag{5.5}$$

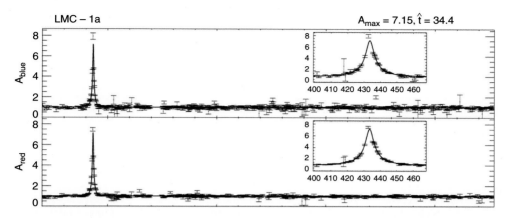

Figure 5.5 A typical microlensing candidate from the MACHO collaboration (Alcock et al., 2000) towards a star in the Large Magellanic Cloud. The star
is magnified for about 34 days (the value $\hat{t}$ given above the plot is defined as
$2t_d$) in both blue and red bands. The color-independence and the single epoch of
magnification are signs that the effect is not due to intrinsic variability of the star.

Therefore, long-duration events correspond to large-mass MACHOs. Fig. 5.5
shows the light curve of a typical microlensing event from the MACHO collaboration (Alcock et al., 2000). The event was one of 15 events detected after observing
the brightnesses of 12 million stars over the course of six years starting in 1992.
The event depicted shows a clear amplification in both red and blue bands only
once during this span, so it is unlikely to be due to a variable star. The inset suggests that the duration of the event is of order 17 days, and the source star is in the
Large Magellanic Cloud, 50 kpc from us. Using the scaling in Eq. (5.5), then, we
would estimate the mass of the lens to be of order $0.1M_{\odot}$ if it were a distance of 5
kpc away.

However, there is clearly a degeneracy between the mass, distance, and velocity
of the lens. Some information can be gleaned from the peak of the magnification,
equal to $1/u_{\min} = \theta_E/\beta_{\min}$, but this depends on the random orientation encoded
in $\beta_{\min}$. Therefore, nothing definitive can be said about these different parameters
from a single event. Rather, one must accumulate events and apply statistics using
three pieces of information:

- Event duration;
- Peak magnification;
- Number of events.

The number of events speaks to how many MACHOs exist in the Galactic halo, so
is of the utmost importance for the dark matter problem.

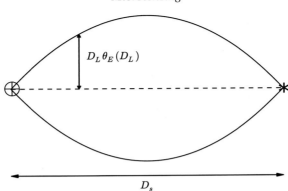

Figure 5.6 Volume between the Earth (on the left) and a distant star (on the right) in which a MACHO will produce an observable event. The 3D region of interest is generated by rotating the solid curves about the dashed axis. If the MACHO is found within this volume, defined by the Einstein radius at every distance D_L, the magnification will be greater than $\mu_T = 1.34$. Note that the figure is completely off-scale with the distance to the star of order tens of kpc, while the radius of the volume is smaller by a factor of $\theta_E \sim 5 \times 10^{-9}$.

Let us calculate the expected number of microlensing events towards a given star a distance D_s from us, assuming that all the dark matter in the Galaxy is in the form of MACHOs. First, we need to define what constitutes an event. Typically, surveys can detect an event with magnification above a threshold μ_T, a little larger than unity. Eq. (5.2) can be inverted to find the resulting threshold for u: the smaller μ_T is, the larger u_T can be. For concreteness, we will set $u_T = 1$ corresponding to the lens and source within an Einstein radius; this corresponds to a peak magnification of $\mu_T = 1.34$.

The geometry is depicted in Fig. 5.6. The expected number of events at any one time is equal to the total number of MACHOs contained within this volume. We will see that this number is quite small, so it can be defined as the *optical depth*, in analogy to the absorption of light due to intervening matter along the line of sight. When the optical depth is large, flux is exponentially suppressed; when small, it gives a sense of the probability for absorption. Here, too, the optical depth gives the probability that a given star will be amplified by a MACHO at a given time. Since it is small, it will also give us a sense of how many stars need to be observed in order to have a realistic chance of observing an event.

What is the chance that a given star at a distance D_s will be magnified by a MACHO? In a given slice along the line of sight with small width dD_L, the area enclosed by the Einstein radius is equal to $\pi D_L^2 \theta_E^2$, so the total volume of this little slice is

$$dV = dD_L \pi D_L^2 \theta_E^2. \tag{5.6}$$

For simplicity, let's assume that all MACHOs have the same mass; then the optical depth from this slice is equal to the number density of MACHOs, n, integrated over each volume slice along the line of sight

$$\tau = \pi \int_0^{D_s} dD_L D_L^2 \, \theta_E^2(M, D_L) \, n(D_L). \tag{5.7}$$

The density of MACHOs depends on the distance D_L. However, what it really depends on is the distance from the center of the Galaxy, as we are assuming a "halo" of them distributed spherically symmetrically around the Galactic center. To simplify the algebra, let's assume that the background star is located at Galactic coordinates $(l = 0, b)$ a distance $R_0 = 8$ kpc from us: i.e., in the bulge in the Galactic center but offset by the declination angle b. Indeed, several surveys searched for events towards the Galactic center. Then, you can show in Exercise (5.3) that the distance of the MACHO from the Galactic center, r, as a function of D_L is equal to

$$r^2 = R_0^2 + D_L^2 - 2R_0 D_L \cos(b). \tag{5.8}$$

Therefore, assuming that the dark matter distribution falls off as $1/r^2$ from the center, the mass density of MACHOs is

$$\rho(D_L) = \rho_0 \frac{R_0^2}{R_0^2 + D_L^2 - 2R_0 D_L \cos(b)} \tag{5.9}$$

where ρ_0 is the local dark matter density, typically estimated to be about $8 \times 10^6 M_\odot$ kpc^{-3}. Dividing this by the MACHO mass M gives the number density:

$$n(D_L) = 8 \times 10^6 \text{ kpc}^{-3} \left(\frac{M_\odot}{M} \right) \frac{R_0^2}{R_0^2 + D_L^2 - 2R_0 D_L \cos(b)}. \tag{5.10}$$

Inserting this into the integral in Eq. (5.7) and setting $D_s = R_0$, since we are considering source stars in the Galactic bulge, we find that the dependence on mass drops out, and

$$\tau = 10^8 \text{ kpc}^{-3} \frac{M G_\odot R_0}{c^2} \int_0^{R_0} dD_L D_L \frac{R_0 - D_L}{R_0^2 + D_L^2 - 2R_0 D_L \cos(b)}. \tag{5.11}$$

A key feature of this integral is that it is independent of the mass of the lens. Of course, this is a bit misleading, as very small masses will lead to very short events that will not be observed; but for masses not too much smaller than $M_\odot$, the independence is real. Therefore, the frequency of events will determine the number density of MACHOs, while the duration and peak magnification will need to be used to extract information about the masses. Plugging in numbers and changing the dummy variable to $x \equiv D_L/R_0$ leads to

$$\tau = 3 \times 10^{-7} \int_0^1 dx \; \frac{x(1-x)}{1 + x^2 - 2x\cos(b)} \qquad \text{Towards Galactic Bulge.} \quad (5.12)$$

The integral is easy to evaluate numerically; for $b = 2.5°$ (roughly the extent of the bulge), it is equal to 2.2, so the dark matter halo produces an optical depth for microlensing towards the Galactic bulge of order 6×10^{-7}.

To detect the signal from even one MACHO in the dark matter halo, then, millions of stars need to be monitored. There have been several experiments that carried out this program and found a rate significantly larger than Eq. (5.12) would suggest (e.g., Popowski et al., 2005). But these bulge events are due mostly to intervening stars. Searches away from the bulge and Galactic disk along lines of sight that contain few stars turn up a fairly low optical depth, consistent with the possibility that only part of the Galactic dark matter is comprised of MACHOs. An example constraint is depicted in Fig. 5.7.

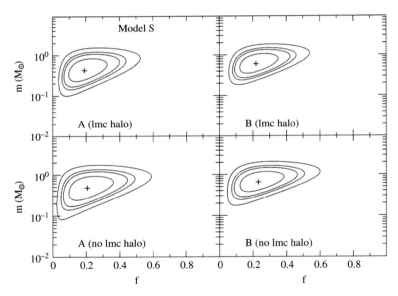

Figure 5.7 Final constraint from six years of data from the MACHO project (Alcock et al., 2000), which monitored 12 million stars in the Large Magellanic Cloud and detected 15 events consistent with microlensing. "A" and "B" denote two different cuts on the data and the bottom panel does not include the possibility of lenses in the LMC itself; the whole analysis is done in the context of a model with dark matter distribution slightly more complex than that given in Eq. (5.9) (the density falls off as $r^2 + a^2$ with $a = 5$ kpc). The contours show the allowed region of parameter space for the mass of the lenses and the fraction of the total halo mass in the Galaxy contained in these lenses. The fraction is constrained to be significantly less than one, so MACHOs cannot be the sole dark matter component.

Although the MACHO searches were initiated in an attempt to discover the nature of the dark matter in the Galaxy, perhaps the most interesting find relates to the nature of the bulge at the center of the Galaxy. Since dust scatters much of the light from the Galactic center, optical surveys do not yield much evidence for a bulge at the center. Infrared surveys are less sensitive to extinction by dust, because the longer wavelength infrared radiation does not scatter off the small dust particles, so some evidence for a bulge emerged from the Diffuse InfraRed Background Experiment (DIRBE) (Dwek et al., 1995), which mapped the sky at wavelengths between 1.25 and 240 microns. But the MACHO searches add even more information. The optical depth constrains the total mass of the central region to be $1.6 \times 10^{10} M_\odot$ and its shape (which is not a spherical bulge, but rather is quite asymmetric). These findings carry immense importance for understanding the formation and structure of the Galaxy, but also provide a crystal-clear illustration of the benefits of lensing: we can learn about structures and their mass even if we cannot see them!

5.3 Exoplanets

The Sun has multiple planets orbiting it, so it has always seemed at least possible that other stars in the Galaxy and in other galaxies would also be the centers of their own solar systems. The famous Drake equation that tries to quantify the likelihood of intelligent life beyond Earth contains the product of several probabilities, one of which (f_p) is the fraction of stars that have planets. Until the 1990s, there was no lower limit on this fraction, while today we know that it is not too far from one.

There are multiple techniques for finding extrasolar planets (or *exoplanets*). To date, the three with the most detections are: Radial Velocity, Transits, and Microlensing. Although our focus here is on the latter, we will walk through all three in this section to place microlensing in context.

Radial Velocity: A planet orbiting a star exerts a gravitational pull on the star causing the star to oscillate with an amplitude of order m/M where m is the smaller mass of the planet and M the mass of the host star. To see this, note that the center of mass of the system, $(m\vec{x} + M\vec{X})/(m + M)$, remains fixed where $\vec{x}$ is the position of the planet and $\vec{X}$ the star. Differentiating leads to an equation for the velocity of the star in terms of the planet's velocity

$$\vec{v}_s \simeq -\frac{m}{M} \vec{v}_p. \tag{5.13}$$

Assuming the planet moves around the star in a circular orbit of radius a, its velocity will be $(MG/a)^{1/2}$. Therefore, if the plane of the orbit is parallel to our line of sight, when the planet is moving away from us in its orbit, the star will be moving towards us with a velocity equal to $m(G/Ma)^{1/2}$. On the opposite side of the star, when the planet is traveling towards us, the star will be moving away at the same speed. At other times during the orbit, the star's radial velocity will be smaller. So the net observable impact of the planet will be an oscillatory radial motion with an amplitude equal to $m(G/Ma)^{1/2}$.

These radial oscillations can be detected by relatively small telescopes as long as they can monitor the star regularly and contain a spectrometer. Initially, the radial velocity technique was the most powerful for identifying exoplanets. Eventually, though, it was used more frequently to confirm candidates identified by small, ground-based telescopes, which use the transit method. Note that the effect is proportional to the mass of the planet, so is more efficient for more massive planets. Also, the induced v_s scales as the square root of the inverse of the distance of the planet from the star. Therefore, the efficiency of the radial velocity technique is highest for planets that have smaller orbits and are more massive. The black points in Figure 5.8 depict planets detected only with the radial velocity method.

Transits: When a planet comes between us and its star, the light observed from the star is blocked, so the flux is diminished by a factor proportional to the area of the planet. Since the mass of a solid planet scales as its radius cubed, this means that the amount by which the flux is diminished – the signal – scales as $m^{2/3}$. This signal is usually small, so multiple observations must be combined to obtain a detection. The signal-to-noise ratio scales as the square root of the number of observations (see, e.g., Exercise (4.8)). In a fixed time, the number of observations that will see the signal is proportional to the fraction of the time that the planet lies between us and the star, a fraction roughly equal to the ratio of the radius of the star to the circumference of the orbit: $R_*/2\pi a$. Therefore, the total signal-to-noise of the transit technique scales as

$$\left(\frac{S}{N}\right)_{\text{transit}} \propto m^{2/3} R_*^{1/2} a^{-1/2}. \tag{5.14}$$

As depicted in Fig. 5.8, it is weighted towards small orbit, large mass planets. The Kepler satellite monitored 150,000 stars and identified 2700 candidates, hundreds of which were confirmed with follow-up radial velocity measurements.

Microlensing: The basic idea of planetary microlensing is that a background star is magnified by a foreground star for a time of order a month, as in the MACHO case discussed above, and for a brief time – less than a day – during that month, the magnification curve is distorted because the image comes very close to the position of the planet. Let's first set up the rough scales. Suppose the primary lens is a solar mass star between us and the Galactic center, say at $D_L = 4$ kpc. If it is magnifying a star in the bulge (8 kpc away), then, from Eq. (1.6), its Einstein radius is 4.8×10^{-9} radians, which corresponds to a physical distance of 4 AU. So

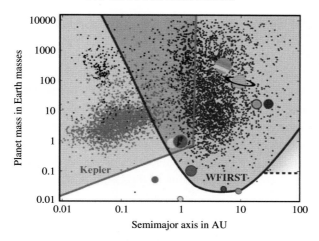

Figure 5.8 Mass and orbital radius of exoplanets discovered by the different techniques (Spergel et al., 2015). The Solar System planets are depicted pictorially. Kepler uses the transit technique and detected planets depicted by the red dots. WFIRST is a proposed satellite mission that will employ microlensing to extend the sensitivity to Earth-mass planets, as indicated by the blue shaded region; blue points are simulated detections with WFIRST. Black points are planets detected using only radial velocities. (See color plates section.)

it is quite likely that – if the star has planets – a planet will be within or near the Einstein radius of its host.

Fig. 5.8 delineates the regions in parameter space for which each technique is best suited. Although there are other techniques (imaging and timing), the vast majority of exoplanets have been discovered with one of the three techniques covered in this section. The transit and radial velocity techniques cover the small orbit region well. As we will see in the next section, the microlensing technique is most effective for planets with larger orbits than the transit or radial velocity technique. A mission such as WFIRST would extend the capability to low mass and large orbits.

5.4 Two Point Mass Lenses

To understand the phenomenology of planetary microlensing, we must generalize the treatment of multiple images and magnification to the case of more than one point mass lens. First we solve for the positions of the images and then, in the next section, use these solutions to obtain the observable: the magnification.

The lens equation for multiple point mass lenses, each with its own Einstein radius $\theta_{E,i}$ at angular position $\vec{\theta}_i$, is

$$\vec{\beta} = \vec{\theta} - \sum_i \theta_{E,i}^2 \frac{\vec{\theta} - \vec{\theta}_i}{|\vec{\theta} - \vec{\theta}_i|^2}. \tag{5.15}$$

There is some ambiguity in the notation here: we are accustomed to using $\vec{\beta}$ for positions in the source plane and $\vec{\theta}$ for positions in the image plane. The multiple lenses here are all denoted with positions $\vec{\theta}_i$, implying that they are image plane positions. But these lenses are themselves *unlensed* so their source plane positions are equal to their image plane positions. Thus, the only unknown here – the position that needs to be solved for – is the image plane position of the background star $\vec{\theta}$. And as this position moves, the flux from the star will change.

The planetary microlensing scenario is one in which, for all practical purposes, the multiple lenses are all at the same distance D_L, since they are in the same solar system. Let's define the Einstein radius of the combined system to be

$$\theta_E^2 \equiv \sum_i \theta_{E,i}^2$$

$$= \frac{MGD_{SL}}{D_L D_S c^2} \tag{5.16}$$

where $M \equiv \sum_i m_i$. Then dividing by this combined Einstein radius, we find

$$\vec{u} = \vec{y} - \sum_i \frac{m_i}{M} \frac{\vec{y} - \vec{y}_i}{|\vec{y} - \vec{y}_i|^2} \tag{5.17}$$

where the source position ($\vec{\beta}$), the image position $\vec{\theta}$, and the lens positions have all been written in units of the Einstein radius: $\vec{u} \equiv \vec{\beta}/\theta_E$; $\vec{y} \equiv \vec{\theta}/\theta_E$ and $\vec{y}_i \equiv \vec{\theta}_i/\theta_E$.

Microlensing by a single planet corresponds to the case where one of the objects (the foreground, host star) contains virtually all of the mass M, while the other (the planet) has much smaller mass m, with ratio $q \equiv m/M$. In this case, the lens equation reduces to

$$\vec{u} \simeq \vec{y} - \frac{\vec{y}}{|\vec{y}|^2} - q\frac{\vec{y} - \vec{y}_p}{|\vec{y} - \vec{y}_p|^2}. \tag{5.18}$$

Here we have chosen the origin to be the position of the foreground heavy object ($\vec{y}_s = 0$) and the subscript denotes the light planet p. Fig. 5.9 shows the basic geometry of this three-body system. We have determined in the MACHO analysis that, fixing the foreground star's position (small black dot at the center in Fig. 5.9), the source is moving across the field of interest over the course of about a month (with its true unlensed locations denoted by the open circles in the figure). What about the planet? The Earth moves around the Sun with a radius of 1 AU in a year. So over the course of a month, an Earth-like planet will change its position relative to its host by less than 10 percent. For practical purposes, then, we can treat the planet as also being fixed. The figure considers two different (fixed) positions for the planet, denoted by + and X. Although both are just outside the Einstein radius,

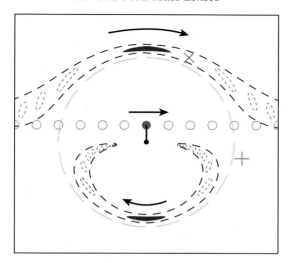

Figure 5.9 Evolution of a planetary microlensing event (Gaudi, 2010). Foreground star is black dot in the center, and open circles denote successive positions of a background star as it moves across the line of sight. The long dashed circle is the Einstein ring of the lens. There are two images of the background star, one outside the Einstein radius (top dotted ovals) and one inside. If the planet lies close to one of these images, then the background star will be further magnified for the brief duration of the overlap. A planet at the position of the cross therefore leads to an observed excess magnification while one at the position of the plus sign does not. (See color plates section.)

we will see that a planet at the position of the X, lying near one of the images of the background star, will have a much bigger observational impact than a planet residing at the position of the $+$.

If we multiplied through by all the denominators in Eq. (5.18), we would get a fifth-order equation for the positions of the images $\vec{y}$, and indeed in some regions of parameter space, there are five images in this system. Regions with five images are separated by a caustic from adjacent regions with three images. Instead of probing the structure of these caustics, though, let's focus on a simple example, which turns out to capture the basic physics underlying planetary microlensing detection. Suppose that the background star, foreground star, and planet are all aligned in the plane perpendicular to our line of sight along the same axis, which we will call the x-axis. Then, $\vec{y}$ and $\vec{y}_p$ have only x-components and the x-component of the lens equation becomes

$$uy(y - y_p) = y^2(y - y_p) - (y - y_p) - qy. \qquad (5.19)$$

Remembering that q is small, let's collect all the other terms on the left into a single third order polynomial. Since q is small, the zeros of that polynomial will be very close to the full solution, which will be only slightly perturbed by the q term. In other words, the lens equation in this special aligned case can be rewritten as

$$P(y) = qy. \tag{5.20}$$

with

$$P(y) \equiv (y - y_p)\left[y^2 - uy - 1\right]. \tag{5.21}$$

One zero of $P(y)$, and therefore one approximate solution, is at $y = y_p$. It takes only a bit of algebra to show that the terms in square brackets can be rearranged to reveal the two roots we found in the single lens cases, as given in Eq. (3.7). That is, the polynomial is simply

$$P(y) = (y - y_p)(y - y_+)(y - y_-). \tag{5.22}$$

Remember that $y_+ \equiv (u/2)(1 + \sqrt{1 + 4/u^2})$ is the location of the brighter image outside the Einstein radius, while $y_- \equiv (u/2)(1 - \sqrt{1 + 4/u^2})$ is the location of the fainter image on the other side of the lens, inside the Einstein radius. So, in the limit that the planet has negligible mass, there are three images of the background star: one outside the Einstein radius, one within the Einstein radius on the opposite side of the primary lens, and a third image at the position of the planet.

The nonzero mass of the planet changes the locations of the images slightly. To find the deviation of the image near y_p, we can set $y = y_p$ everywhere except in the $(y - y_p)$ factor. Therefore, the polynomial $P(y) \simeq (y - y_p)(y_p + -y_-)(y_p - y_+)$. Then, setting this equal to the right-hand side of Eq. (5.20), qy, leads to a shift of

$$y - y_p \simeq \frac{qy_p}{(y_p - y_+)(y_p - y_-)}. \tag{5.23}$$

Fig. 5.10 depicts the effect of the planet's mass: the image previously located at the position of the planet, y_p, moves towards the larger mass (at $y_s = 0$) because the mass of the planet in this case acts in the opposite direction as that of the foreground star. The image of the background star moves to the right of y_+ farther away from the lens. This is what we would expect: the planet essentially adds to the mass of the foreground star, producing a slightly larger deflection. This effect is of order m/M, quite small. A similar effect holds for the opposite side image, which is perturbed farther away from the primary lens.

These small shifts will in general go undetected; indeed, the images are not resolvable. So, in order to detect the influence of the planet, we need to observe time-varying magnification.

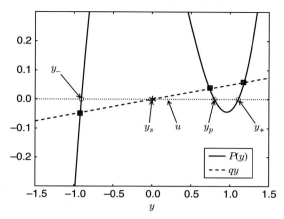

Figure 5.10 Positions of images produced by two point masses. Open circles give the position of the images in the limit that the lighter mass is infinitely small. When q, the ratio of the masses, becomes non-negligible, the images lie not where the polynomial defined in Eq. (5.21) is zero, but rather when it crosses the line qy. Therefore, the images move and appear at the positions of the solid squares. The foreground lensing star is at the origin here (y_s); the planet is at y_p; and the unlensed position of the background star is u (in this case chosen to be equal to 0.2).

5.5 Exoplanet Microlensing: Magnification

The magnification of an image is given (Eq. (4.1)) by the inverse of the determinant $|\partial\beta_i/\partial\theta_j|$, or with the Einstein radius-normalized variables we have been using: $|\partial u_i/\partial y_j|$. Differentiating Eq. (5.18) with respect to $\vec{y}$ leads to a 2×2 matrix (the unit matrix plus the distortion tensor of Eq. (4.4)):

$$
\begin{pmatrix}
1 + \frac{y_x^2 - y_y^2}{y^4} + q\,\frac{(y-y_p)_x^2 - (y-y_p)_y^2}{|\vec{y}-\vec{y}_p|^4} & \frac{2y_x y_y}{y^4} + q\,\frac{(y-y_p)_x(y-y_p)_y}{|\vec{y}-\vec{y}_p|^4} \\[2mm]
\frac{2y_x y_y}{y^4} + q\,\frac{(y-y_p)_x(y-y_p)_y}{|\vec{y}-\vec{y}_p|^4} & 1 + \frac{y_y^2 - y_x^2}{y^4} + q\,\frac{(y-y_p)_y^2 - (y-y_p)_x^2}{|\vec{y}-\vec{y}_p|^4}
\end{pmatrix}.
\tag{5.24}
$$

Here I have simplified by placing the lensing star at the origin ($\vec{y}_s = 0$), but the notation is still confusing: the image is at the 2D position $\vec{y}$, which has both an x and a y component. In our simple case, where the images lie along a single axis, all the y-components can be set to zero, and the magnification then becomes

$$
\mu = \left[\det \begin{pmatrix} 1 + \frac{1}{y^2} + \frac{q}{(y-y_p)^2} & 0 \\[2mm] 0 & 1 - \frac{1}{y^2} - \frac{q}{(y-y_p)^2} \end{pmatrix} \right]^{-1}
$$

$$
= \left[1 - \left(\frac{1}{y^2} + \frac{q}{(y-y_p)^2} \right)^2 \right]^{-1}.
\tag{5.25}
$$

We recapture the single-lens result of Eq. (4.12) when the planet mass is set to zero: $\mu \to \mu_0 \equiv |1 - y^{-4}|^{-1}$. Here μ_0 is the magnification curve in the absence of the planet; that is, it is the typical month-scale time-varying magnification shown in Fig. 5.5.

What is special here is the perturbation: the effect of the planet on the magnification. To extract this effect, let us rewrite the magnification as

$$\mu = \frac{1}{1 - \frac{1}{y^4}\left[1 + \frac{qy^2}{(y-y_p)^2}\right]^2}. \tag{5.26}$$

The total magnification is the sum of that from each of the three images, the one perturbed slightly from y_p and the two perturbed from $y_\pm$. The first of these contributes little to the magnification. To see this, note that Eq. (5.23) demonstrates that $y - y_p$ for this image is proportional to q. Therefore the last term in the denominator of Eq. (5.26), $q^2/(y - y_p)^4$, scales as $1/q^2$; i.e., it is very large. Therefore, this image is highly de-magnified. Excess magnification due to the presence of the planet must come from one of the other two images, those near $y_\pm$. For these images, we can approximate the square of the square brackets by dropping the $q^2 y^4/(y - y_p)^4$ term so that

$$\mu \simeq \frac{1}{1 - \frac{1}{y^4}\left[1 + \frac{2qy^2}{(y-y_p)^2}\right]}. \tag{5.27}$$

The first two terms in the denominator are now equal to the inverse of the magnification in the absence of the planet, so,

$$\mu \simeq \frac{1}{\mu_0^{-1} - \frac{2q}{y^2(y-y_p)^2}}$$

$$\simeq \mu_0 \left[1 + \mu_0 \frac{2q}{y^2(y - y_p)^2}\right]. \tag{5.28}$$

Again, here, only the linear term in the Taylor expansion in q has been retained. This formula holds the key to exoplanet detection from microlensing, so we now turn to its implications.

Consider the excess magnification due to the mass of the planet: call it $\Delta\mu_p$. From Eq. (5.28), we see that

$$\Delta\mu_p = \frac{2\mu_0^2 q}{y^2(y - y_p)^2}. \tag{5.29}$$

The denominator here tells the story: if the planet is close to the position of one of the images, then the magnification will be large. Fig. 5.9 shows this with a cartoon. As the source position moves across the primary lens, the two image positions –

one outside the Einstein ring and the other inside – follow along. If the planet lies near one of the images (e.g., if it were located at the X), then for a short time the $y - y_p$ term in the denominator would be small and a brief epoch of excess magnification would occur; if it lies far from the images (e.g., if it were located near the plus sign), there would be little enhancement. The dashed curves along the image trajectories give the sense of how close the planet needs to be to one of the images. Let's quantify this. Suppose that a given survey is sensitive to excess magnifications above some threshold $\Delta\mu_T$ (typically of order one). This translates to the requirement that

$$|y - y_p| < \Delta y \equiv \frac{\mu_0}{y} \left(\frac{2q}{\Delta\mu_T}\right)^{1/2}. \tag{5.30}$$

This angular distance scales as $q^{1/2}$, so even relatively small mass planets can significantly impact the magnification curves. As an example, even an Earth-mass planet with $q = 2 \times 10^{-6}$ would produce a detectable signal on top of a light curve with $\mu_0 = 10$ (a fairly common event) if $\Delta y \lesssim 0.02$. That is, the magnification due to the planet would be observable above the nominal magnification curve due to the host star as long as the planet was close to the background stellar image.

How often will this happen? How often will the planet be close enough to the stellar image to produce an observable microlensing signature? The answer depends on how often the planet happens to lie within the semi-annulus traced out by the dashed lines in the upper region of Fig. 5.9. Approximating this region as a half of a circular annulus with radius y_+ and width $2\Delta y$, we find that the small area of detectability is

$$\Delta A \simeq 2\pi\, \Delta y y_+. \tag{5.31}$$

The fraction of planets that would produce a signal then corresponds to the ratio of ΔA to the total area (roughly πy_+^2). So, the detection probability is roughly

$$\text{Prob} \sim 2\mu_0 (2q)^{1/2}, \tag{5.32}$$

again proportional to the square root of the planet star mass ratio. For an Earth-mass planet orbiting a solar mass star, this corresponds to a probability of order 0.8 percent during a microlensing event with $\mu_0 = 2$. This simple estimate holds only for the particular alignment discussed above. More generally, the probability scales as $q^{5/8}$ and is closer to 0.4 percent for an Earth mass planet.

We can also estimate the duration of the blip in the magnification curve due to the planet. Since the planet remains fixed over the time scales of interest, the blip persists as long as

$$|y - y_p| \lesssim \mu_0 q^{1/2}. \tag{5.33}$$

But near the planet, $y \simeq y_p + v_\perp t/(D_L\theta_E)$. Therefore, the event remains detectable for a duration

$$\Delta t \sim \mu_0 q^{1/2} \left(\frac{D_L\theta_E}{v_\perp}\right). \tag{5.34}$$

The term in parentheses is the duration of the primary microlensing event, the time it takes for the image to traverse the Einstein radius. So the surge in magnification due to planetary effects will persist for roughly a percent of the primary event duration. Instead of months, planetary microlensing hunters need to monitor fluxes several times a day.

Fig. 5.11 shows a light curve from a detection of an exoplanet via microlensing (Skowron et al., 2015). The main microlensing occurred over the course of a few months and was first observed by the Optical Gravitational Lensing Experiment (OGLE) collaboration, which used the 1.3m Warsaw University Telescope deployed in Chile. To detect the effect of planets, a larger collaboration had been formed to quickly and repeatedly follow up a microlensing event with a suite of telescopes. The observations shown in Fig. 5.11 were made from small telescopes around the world. There is clear evidence for a blip that lasted for several days. That bump in the light curve is best fit with a model in which a planet with a mass a little smaller than that of Jupiter orbits the host star at a separation of 2 AU.

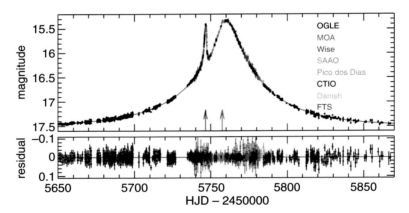

Figure 5.11 Light curve of a star in the Galactic bulge that is magnified over the course of a few months by an intervening star and planet (Skowron et al., 2015). The initial event was detected by the OGLE collaboration but then followed up with multiple observations at several different telescopes listed in the figure. The blip at Heliocentric Julian Date (HJD) $\simeq$ 5746 days is due to the Jupiter-sized planet. The four-day blip induced by the planet is well fit by the solid curve, as seen by the difference between the data and the model in the bottom panel. (See color plates section.)

In addition to opening up the possibility of detecting Earth-like planets, microlensing has already taught us something fascinating about our Galaxy (that presumably holds for other galaxies as well), and more generally our place in the universe. Roughly one in a million stars produces a microlensing signal that can be observed by OGLE or similar surveys. As we saw above, not all of these microlensing events would lead to an observable planet detection even if the star in question had a planet and even if the network of follow-up telescopes provided perfect coverage. To estimate how many stars have planets, one must obtain the efficiency of a given survey for detecting planets. Several groups have done this and concluded that the number of planets detected (e.g., about 10 during the years 2002–2007) suggests that on average every star in the Milky Way hosts at least one planet. We are not only not alone: we have a lot of company! There are likely hundreds of billions of planets in our Galaxy, and likely as many in each one of the other billions of galaxies in the observable universe.

Suggested Reading

The idea of searching for MACHOs using microlensing was first introduced by Paczynski (1986); the treatment in §5.2 follows the very clear paper by Griest (1991). The large optical depth towards the bulge is described in Popowski et al. (2005); an explanation of the contribution of stars to this optical depth is given by Han and Gould (2003). Implications for the structure of the Galactic bulge (the "peanut at the center of the Galaxy") are reviewed in Minniti and Zoccali (2008).

There are many excellent reviews of the rapid developments and various techniques of detecting exoplanets. Listing the number of detected exoplanets in a book would be a bit like a newspaper going to print listing how many points Michael Jordan has scored in the first quarter. So more relevant are reviews of the techniques, key issues, and projects. A short article by Bhattacharjee and Clery (2013) gives a nice overview of all these. More technical, but still very clear, is the chapter by Wright and Gaudi (2013). Planetary microlensing was proposed and explained clearly in Gould and Loeb (1992); Gaudi (2010) has a more recent detailed explanation of the physics.

Exercises

5.1 The luminosity from stars in the disk falls off exponentially as the 2D radial distance from the center of the Galaxy increases:

$$L(R) = L_* e^{-R/R_*} \qquad (5.35)$$

with $R_* \simeq 3$ kpc. Determine L_* by requiring that the total stellar luminosity is equal to $2 \times 10^{10} L_\odot$. If every star had mass and luminosity equal to that of the Sun, what would the rotation curve look like at $R \gg R_*$?

5.2   Assuming the parameters given in Exercise (5.1) and the indication from Fig. 5.4 that the rotation velocity at the position of the Sun (a distance $R_0 \simeq 8$ kpc from the Galactic center) is of order 200 km/sec, determine the value of ρ_0, the dark matter density parameter, assuming a profile

$$\rho_{DM} = \rho_0 \left(\frac{R_0}{r} \right)^2. \tag{5.36}$$

5.3   Compute the distance r from the Galactic center to a MACHO (labeled 'M' in Fig. 5.12) as a function of D_L and Galactic latitude b. Show that it is given by Eq. (5.8).

5.4   Determine the optical depth for microlensing towards a star in the Galactic bulge due to dark matter as a function of Galactic latitude b (with $l = 0$) by numerically integrating Eq. (5.12).

5.5   Calculate the optical depth for a star in the bulge to be microlensed by a star along the line of sight. Take the stellar density to fall off exponentially with scale length $R_D = 3.5$ kpc and take the stellar density near the Sun to be $0.044 M_\odot/\text{pc}^3$.

5.6   Determine the planet mass that produces oscillations in the stellar velocity with amplitude of 1 m/s as a function of the orbital radius of the planet.

5.7   Estimate the contamination to planetary microlensing (towards stars in the bulge) from stars along the line of sight that happen to lie near the Einstein ring of the primary lensing star.

5.8   Calculate the amplitude of the radial oscillations (in m/s) of a star with solar mass orbited by a planet with Earth mass and period. Repeat for Jupiter mass and period.

5.9   After Eq. (5.33), the claim is made that $y \simeq y_p + v_\perp t/(D_L \theta_E)$. Show that this is true by Taylor expanding the expression for $y_+(u)$ as u changes with time.

5.10 Compute the excess magnification induced by a planet an angular distance Δ from the image (of a lensed star) if – in the plane perpendicular to the line

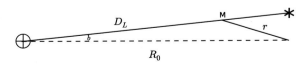

Figure 5.12 MACHO located a distance D_L from us at Galactic latitude b with $l = 0$.

of sight – the line connecting the planet to the image is perpendicular to the line connecting the primary lens to the image. Treat Δ as small compared to the other angular scales. Note: in the text, we computed the excess magnification when all three – primary lens, image, and planet – lie along the same line.

6

Weak Lensing: Galaxy Shapes

Until now, we have been implicitly assuming that the objects being lensed are point sources. We now transition to the more general case when the background objects are extended. Fig. 1.7 shows a dramatic example of a background extended galaxy distorted by a foreground lens. The image is stretched (or *sheared*) and brightened. In terms of the distortion tensor,

$$\Psi_{ij} = \begin{pmatrix} \kappa + \gamma_1 & \gamma_2 \\ \gamma_2 & \kappa - \gamma_1 \end{pmatrix}, \tag{6.1}$$

the shear is quantified by the two components, $\gamma_{1,2}$, while the dilation or magnification is described by the convergence κ. The components of shear are directly related to Φ via Eq. (4.5), so measuring them opens up the possibility of inferring information about the projected gravitational potential along the line of sight. Here we discuss one of the most promising ways of measuring these two components of shear: measuring the shapes of background galaxies. We will see that these measurements are difficult, so please keep in mind while reading this chapter that the reward for measuring shapes is gleaning direct information about the mass distribution in the universe.

6.1 The Effect of Shear

Lensing leads to dilation and stretching of images. We saw in Chapter 4 the way that κ dilates an image; now let us consider the effects of shear. For simplicity, let's examine the effect of the components of the distortion tensor on the image of a background circular object (think galaxy) with angular radius β_0.

First consider an unlensed circular object distorted by Ψ_{ij} with $\kappa = \gamma_2 = 0$ and γ_1 small but positive. In this case, the lens equation dictates that β_1 depends only on θ_1, and β_2 depends only on θ_2. The ordinary differential equations determining the location of the image, then, are

Figure 6.1 The effects of elements of the distortion tensor on an initially circular background object (the same sized dashed circle in each panel). The first row shows the effect of $\gamma_1 \neq 0$; positive γ_1 produces an elongated ellipse along the x-axis, while for negative γ_1 the elongation is aligned with the y-axis. The bottom row shows the effect of γ_2. In all images, the amplitude of the nonzero component of the distortion tensor (either γ_1 or γ_2) has been set to 0.2.

$$\frac{d\beta_1}{d\theta_1} = 1 - \gamma_1$$

$$\frac{d\beta_2}{d\theta_2} = 1 + \gamma_1. \tag{6.2}$$

Solving and then adding the two components in quadrature leads to

$$\beta_1^2 + \beta_2^2 = (1 - \gamma_1)^2 \theta_1^2 + (1 + \gamma_1)^2 \theta_2^2. \tag{6.3}$$

But the unlensed source is assumed to be a circle with radius β_0, so the left-hand side is a constant β_0^2 and the resulting equation

$$(1 - \gamma_1)^2 \theta_1^2 + (1 + \gamma_1)^2 \theta_2^2 = \beta_0^2 \tag{6.4}$$

tells us that the image is an ellipse. As depicted in Fig. 6.1, the major axis is aligned with the x-axis with radius $\beta_0/(1 - \gamma_1)\sqrt{2}$ while the minor axis is aligned with the y-axis and has radius $\beta_0/(1 + \gamma_1)\sqrt{2}$.

Finally, when γ_2 is the only nonzero component of the distortion tensor, you can show in Exercise (6.1) that

$$\beta_0^2 = (\theta_1 - \gamma_2\theta_2)^2 + (\theta_2 - \gamma_2\theta_1)^2. \tag{6.5}$$

For γ_2 positive, this is the equation for an ellipse with long axis aligned with the x-axis rotated counterclockwise about the x-axis by an angle of $45°$, as depicted in Fig. 6.1.

6.2 Ideal Relation Between Shear and Ellipticity

The key to inferring the shear is to measure the shapes of background galaxies. These shapes, as quantified by the ellipticities, are affected by the shear. The simplest way to see this is to compute the two components of ellipticity of a background galaxy that would appear circular in the absence of lensing. The ellipticity can be quantified in terms of moments of the surface brightness. In particular, if we define

$$q_{ij} \equiv \int d^2\theta \, S^{\text{obs}}(\vec{\theta})\theta_i\theta_j, \tag{6.6}$$

then the two components of ellipticity are

$$\epsilon_1 \equiv \frac{q_{11} - q_{22}}{q_{11} + q_{22}}$$

$$\epsilon_2 \equiv \frac{2q_{12}}{q_{11} + q_{22}}. \tag{6.7}$$

For example, if the object is elongated along the x-axis, then q_{11} will be larger than q_{22} and ϵ_1 will be positive. Indeed, you can quickly see that the orientations of ϵ_1 and ϵ_2 follow the two components of shear depicted in Fig. 6.1.

In this simple example, we are assuming that the unlensed object is circular, so that $S(\vec{\beta}) = S\left(|\vec{\beta}|\right)$. Here, the origin is assumed to be at the center of the image, so the brightness in the source plane depends only on the magnitude of the distance from the origin, $|\vec{\beta}|$. It would be easiest then to compute the moments defined in Eq. (6.6) by integrating over $d^2\beta$ in the source plane instead of the image positions $d^2\theta$. There are two steps needed to make this change in dummy variables. First, the Jacobian $|\partial\beta_i/\partial\theta_j|$ must be computed so that

$$d^2\theta \rightarrow \frac{d^2\beta}{|\partial\beta_i/\partial\theta_j|}. \tag{6.8}$$

The Jacobian is the determinant of the 2×2 matrix with elements

$$A_{ij} \equiv \partial\beta_i/\partial\theta_j = \delta_{ij} - \Psi_{ij}. \tag{6.9}$$

However, we need not calculate it as it does not vary over the extent of typical background objects, so it can be removed from the integral. Since it appears in both the numerator and the denominator of Eq. (6.7), the Jacobian cancels out. The second shift is to change the image positions θ_i to the source positions β_i. Again, we can invoke the fact that over the extent of the object Ψ_{ij} is roughly constant, so the derivative of the lens equation can be integrated to yield

$$\beta_i = A_{ij}\theta_j. \tag{6.10}$$

So both occurrences of θ_i in the integrand of Eq. (6.6) that defines the moments can be replaced by $(A^{-1})_{ii'}\beta_{i'}$. We are left with

$$q_{ij} = \mathcal{N}(A^{-1})_{ii'}(A^{-1})_{jj'}\int d^2\beta \, S(\beta)\,\beta_{i'}\beta_{j'} \tag{6.11}$$

where $\mathcal{N}$ is a normalization factor that will drop out when the ellipticities are computed.

We will shortly compute these moments in terms of the components of the distortion tensor, but it is worth pausing and justifying an assumption we have now made two times: the shear is constant over the extent of the image. We will typically be thinking of galaxies as the objects whose shapes are distorted, and the typical size of a distant galaxy is of order a few arcseconds (see Exercise (6.4)). Meanwhile the shear field produced by an object like a foreground galaxy, galaxy cluster, or large scale structure in general is typically measured on much larger scales, of order arcminutes or larger. So the assumption that the distortion tensor, and therefore A_{ij}, is constant over the extent of the background object is a very good one.

We can calculate the inverse of the 2×2 matrix A in terms of the elements of the distortion tensor in Eq. (6.1):

$$A^{-1} = \frac{1}{(1 - \kappa)^2 - \gamma_1^2 - \gamma_2^2} \begin{pmatrix} 1 - \kappa + \gamma_1 & \gamma_2 \\ \gamma_2 & 1 - \kappa - \gamma_1 \end{pmatrix}. \tag{6.12}$$

If the background galaxy is circular, then the integral in Eq. (6.11) is proportional to δ_{ij} since there is no preferred direction. Again, the value of this integral drops out of the ratios that form the ellipticities, so the expression for ϵ_2 for example becomes

$$\epsilon_2 = \frac{2(A^{-1})_{1i} (A^{-1})_{2i}}{(A^{-1})_{1i} (A^{-1})_{1i} + (A^{-1})_{2i} (A^{-1})_{2i}}. \tag{6.13}$$

The denominator of A^{-1} in Eq. (6.12), the determinant of the matrix A, drops out of the ellipticity. Carrying through the algebra leads to

$$\epsilon_2 = 2 \frac{(1 - \kappa + \gamma_1)\gamma_2 + \gamma_2(1 - \kappa - \gamma_1)}{(1 - \kappa + \gamma_1)^2 + \gamma_2^2 + (1 - \kappa - \gamma_1)^2 + \gamma_2^2}$$
$$= 2\gamma_2 \frac{(1 - \kappa)}{(1 - \kappa)^2 + \gamma_1^2 + \gamma_2^2}. \tag{6.14}$$

Similarly, you can carry out the calculation for ϵ_1 (Exercise (6.5)) and show that

$$\epsilon_1 = 2\gamma_1 \frac{(1 - \kappa)}{(1 - \kappa)^2 + \gamma_1^2 + \gamma_2^2}. \tag{6.15}$$

Equations (6.14) and (6.15) are the starting point of our foray into the world of shapes: they motivate the measurement of the ellipticity, because the two components of ellipticity are clearly related to the analogous components of shear, and these in turn carry information about the gravitational potential along the line of sight. Be forewarned that the factor of 2 in front of both is a convention that follows from the definitions in Eq. (6.7), a convention that is not universally adopted (see

Exercise (6.6) for a convention that eliminates the factor of 2). Finally, note that, in both cases, the ellipticity can be rewritten as

$$\epsilon_i = \frac{2\gamma_i}{1 - \kappa} \frac{1}{1 - [\gamma_1^2 + \gamma_2^2]/(1 - \kappa)^2} \tag{6.16}$$

where the factor in front, $\gamma_i/(1 + \kappa)$, is called the *reduced shear* and is the relevant quantity that influences the galaxy shapes. The second factor on the right here is rarely important, so can usually be neglected, but note that it too depends on the square of the reduced shears. The conclusion is that measurements of the shapes of background objects yield information only about the reduced shear. If all elements of the distortion tensor are small, then there is little difference between the reduced shear and the shear itself, and indeed this is the domain of *weak lensing*. However, even for systems that are weakly lensed, the convergence can at times be large enough that the reduced shear must be used.

6.3 Weighting

Consider the left and right panels of Fig. 6.2. The left panel shows a simulated image of a galaxy with appreciable ellipticity oriented along the x-axis ($\epsilon_1 > 0$). Many of the pixels around the image, though, are dark, so they convey no information. The question arises, then: which pixels should be used when computing the moments defined in Eq. (6.6)? Suppose one were to use only the information contained within the innermost black circle. By eye, we can see that the inferred ellipticity will be very small since there is little asymmetry within this small inner region. Indeed, the estimated ellipticity if only the region within this circle were used to compute the moments would be only 10 percent of the true ellipticity. Moving outwards, the middle circle traced by the dashed white line clearly encompasses a region that is asymmetric, and the estimated ellipticity from within this region would be over 50 percent of the true ellipticity. Moving to the outermost circle gets 90 percent. The clear lesson is that, in order to measure the ellipticity, it pays to use as much of the area as possible, at least in the noiseless case.

Now turn to the right panel in Fig. 6.2, where noise has been added in. The lesson learned above holds, but only up to a point. Going beyond the middle circle actually leads to a *decrease* in the estimate of the ellipticity, as shown in Fig. 6.3. The noise pollutes the signal enough that the inferred ellipticity goes down as one includes a larger region.

There is tension, therefore, between two competing needs: we want to capture as much of the signal as possible, so want to include as much of the area as possible, but including more area blindly often leads to including more noise, thereby

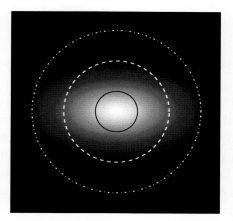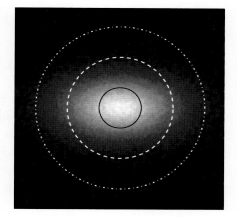

Figure 6.2 *Left Panel*: Simulated noiseless image of galaxy with noticeable ellipticity oriented along the x-axis. Including only pixels within the inner circle in the measurement of the moments will lead to an estimate of the ellipticity that is only about 10 percent of the true ellipticity; middle dashed circle about 50 percent; even the white outermost circle misses about 10 percent of the ellipticity. In the noiseless case, then, it is best to include information as far as possible from the center of the image. *Right panel*: Same image, this time with noise. Now pixels within the middle circle capture a little less than 50 percent of the ellipticity, while the estimated ellipticity is *smaller* if one uses all the pixels within the outermost circle. Noise forces one to use a more clever weighting function.

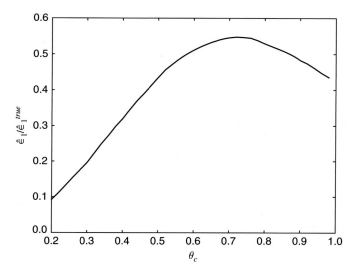

Figure 6.3 For the image with noise in the right panel of Fig. 6.2, ratio of the inferred ellipticity $\hat{\epsilon}_1$ to the true ellipticity as a function of the angular radius θ_c of the circle within which the moments are computed. Here, $\theta_c = 1$ corresponds to half the length of the image in Fig. 6.2. As one moves to larger radii, the noise pollutes the signal, so the inferred ellipticity decreases.

diluting the signal. The solution is to introduce a *weighting function*, so that the moments are computed via

$$Q_{ij} \equiv \int d^2\theta \, S^{\text{obs}}(\vec{\theta}) \, W(\vec{\theta}) \theta_i \theta_j \tag{6.17}$$

instead of the unweighted versions in Eq. (6.6). The visual exercise above was a primitive form of weighting: we included only the region within one of the circles with radius θ_c, which corresponds to setting the weighting function equal to $W(\vec{\theta}) = \Theta(\theta_c - |\vec{\theta}|)$, where Θ is the Heavyside step function equal to one if its argument is positive and zero otherwise. Different groups use different weighting functions; indeed, a weighting function is one key choice that must be made when defining an algorithm to extract shear from galaxy images.

The two components of ellipticity are now defined as in Eq. (6.7), but with $q_{ij} \rightarrow Q_{ij}$. It will prove useful to define also the sum of the diagonal moments, or the trace, as

$$T \equiv Q_{11} + Q_{22}. \tag{6.18}$$

The ellipticities of the galaxies can then be expressed as a ratio of the moments and T. Explicitly,

$$\epsilon_1 \equiv \frac{Q_{11} - Q_{22}}{T}$$

$$\epsilon_2 \equiv \frac{2Q_{12}}{T}. \tag{6.19}$$

6.4 Ellipticity and Shear with Weighting and Intrinsic Ellipticity[1]

We are now in a position to move beyond the ideal ellipticity–shear relation of Eq. (6.16) and include two necessary complications: (i) the weighting of moments and (ii) the fact that the unlensed galaxies are not always circular. This second effect, called *intrinsic ellipticity*, is obvious: most galaxies we see in the sky, even the ones close to us, that are by and large unlensed, do not appear circular. Intrinsic ellipticity means that a measurement of ϵ_1, ϵ_2 does not immediately translate into a measurement of shear. Indeed, one might wonder how shear can be inferred at all if every observed ellipticity could have nothing to do with the shear field, i.e., could be intrinsic. The short, qualitative answer is that the intrinsic ellipticities of galaxies are (more or less) random, so if we average the ellipticities of many galaxies in the same region on the sky, the mean intrinsic ellipticity will be zero, while the ellipticity due to shear will be the same for each galaxy. This is a classic case of the signal (ellipticity due to shear) emerging from the noise (intrinsic ellipticity) since the noise averages to zero over many measurements while the signal persists.

---

[1] Sections 6.4 and 6.5 are quite technical and probably should be skipped or skimmed on a first reading.

Clearly, then, surveys will be most powerful in measuring the shear if they image many background galaxies.

The goal of this section, then, is to relate the observed ellipticities as defined in Eq. (6.19) to the two components of shear defined in Eq. (6.1) in the presence of weighting and intrinsic ellipticity. The starting point is the relation between the observed intensity, in the image plane, and the unlensed intensity in the source plane:

$$\mathcal{S}^{obs}(\vec{\theta}) = \mathcal{S}(\vec{\beta}) = \mathcal{S}(\vec{\theta} - \vec{\alpha}). \tag{6.20}$$

That is, the photons we see at position $\vec{\theta}$ come from source position $\vec{\beta}$, which the lens equation relates to $\vec{\theta}$ and the deflection angle $\vec{\alpha}$.

There are several ways to proceed, but perhaps the simplest is to Taylor expand the surface brightness around the small deflection angle α, but there is a subtlety. If every photon is deflected by the same amount – that is, if α is constant – the object's shape will not be distorted at all. So we will indeed Taylor expand but drop the constant α piece. We will assume that the distortions (the shear and convergence) are small, and so will be a bit less ambitious than in §6.2, where we were able to derive the full nonlinear relation between ellipticity and the elements of the distortion matrix in the ideal case. Here, we are aiming to capture only the linear term, but in the context of accounting for weighting and intrinsic ellipticity. In this case, to leading order the surface brightness becomes

$$\mathcal{S}^{obs}(\vec{\theta}) \simeq \mathcal{S}(\vec{\theta}) - \frac{\partial \mathcal{S}}{\partial \theta_l} \alpha_l$$

$$\simeq \mathcal{S}(\vec{\theta}) - \frac{\partial \mathcal{S}}{\partial \theta_l} \Psi_{lm} \theta_m. \tag{6.21}$$

To obtain the second line, we expand $\alpha_l(\vec{\theta}) \simeq \text{Constant} + (\partial \alpha_l/\partial \theta_m) \theta_m$, drop the constant term, and then identify the partial derivative as the distortion tensor. The last line of Eq. (6.21) gives us a sense of where we are headed: as we take moments of the observed surface brightness, we will indeed extract terms linear in γ_1, γ_2, the shear components of the distortion tensor, but these will be multiplied by rather complicated integrals over the the weighted moments. The complexity will play itself out over the next few paragraphs, but the key idea – that the observed shape encodes information about the shear – is clear in Eq. (6.21).

We can now take the moments of the surface brightness as defined in Eq. (6.17):

$$Q_{ij}^{obs} \simeq Q_{ij} - \Psi_{lm} \int d^2\theta \, \frac{\partial \mathcal{S}(\vec{\theta})}{\partial \theta_l} \, W(\vec{\theta})\theta_i\theta_j\theta_m. \tag{6.22}$$

Here, we make the usual assumption that the distortion tensor does not vary over the extent of the image so can be removed from the integral. The first term on the right will lead to the intrinsic ellipticity. The second term contains the unlensed surface brightness S, which is of course unknown. However, the difference between the lensed and unlensed S is proportional to the distortion tensor; since this term already has one factor of Ψ and we are working only to linear order, we can ignore the distinction between S^{obs} and S in this term. From now on, all the factors of S on the right can be evaluated using the observed surface brightness.

Integrating by parts leads to a compact expression for the change in the moments:

$$Q_{ij}^{obs} = Q_{ij} + \Psi_{lm} Z_{lmij} \tag{6.23}$$

where

$$Z_{lmij} = \int d^2\theta S(\vec{\theta}) \frac{\partial}{\partial \theta^l} [\theta_i \theta_j \theta_m W(\theta)]$$

$$= \int d^2\theta S(\vec{\theta}) \left(W [\delta_{il}\theta_j\theta_m + \delta_{lm}\theta_j\theta_i + \delta_{jl}\theta_i\theta_m] + 2W'\theta_i\theta_j\theta_l\theta_m \right). \tag{6.24}$$

Here W' is the derivative of W with respect to θ^2, so if W were a Gaussian with mean zero and width σ, W' would be $-W/2\sigma^2$. We will keep things a bit more general by leaving in W' instead of $-W/2\sigma^2$, but note that the weighting function has been implicitly chosen to depend only on the angular distance θ: $W(\vec{\theta}) = W(|\vec{\theta}|)$. This is not obviously the best thing to do. Looking at Fig. 6.2, we can imagine that a better choice of weighting function is one that traces the contours of the galaxy shape. An elliptical weighting function then would minimize the noise and maximize the signal. Indeed, most of the work on galaxy shapes that uses this moment technique does employ weighting functions that are iterative, that rely on initial estimates of the shape and then refine the weighting function to obtain optimal estimates of the shapes. For our initial foray into the world of shape estimation, let's stick to the simple weighting function that is symmetric.

Eq. (6.23) illustrates that the observed moments are equal to the unlensed moments (the first term on the right) plus a correction term due to lensing. Since the ellipticity is a function of the moments, this first term propagates forward to become the intrinsic ellipticity. The term proportional to the distortion tensor must now be massaged. As we will see, there is a piece in Z_{lmij} that is independent of the shape of the background galaxy, so is similar to the ideal cases in Eq. (6.16), but there is also another piece that depends on the intrinsic ellipticity. That is, the measured ellipticity has one term equal to the intrinsic ellipticity; one proportional to the shear; and another that depends on both.

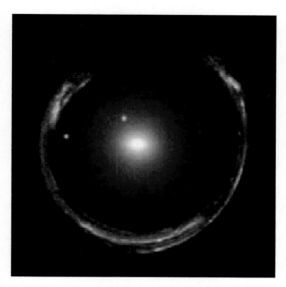

Figure 1.5 Einstein ring first observed by the Sloan Digital Sky Survey and then followed up with the Hubble Space Telescope. Foreground lens is a galaxy at the center, while the background object is almost perfectly aligned so is seen as a ring (Credit: ESA/Hubble & NASA).

Figure 1.7 The shapes of background galaxies made highly elliptical by the foreground Abell cluster (Credit: Gravitational lensing in galaxy cluster Abell 2218: NASA, A. Fruchter and the ERO Team, STScI).

Figure 1.10 Map of anisotropies in the temperature of the CMB by the Planck satellite. The mean temperature, $T = 2.725K$, has been subtracted off so the red and blue regions represent spots only slightly hotter and colder than average, with the typical excess temperature of these spots $\sim 80\mu K$, and the typical size degree-scale (Credit: ESA).

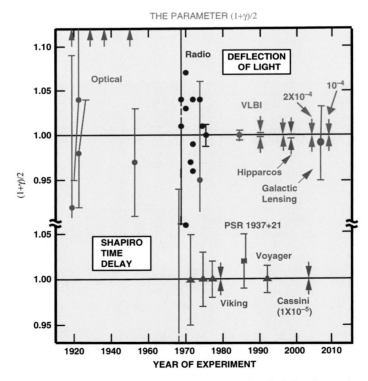

Figure 2.3 Constraints on deviations from general relativity from time delay experiments from Will (2014). The bottom half of the figure depicts constraints from detections of the time delay signal, such that the general relativity prediction corresponds to amplitude $(1 + \gamma)/2 = 1$, as discussed in the text. The top half shows constraints from the deflection of light, such as the first detection during the Solar eclipse of 1919.

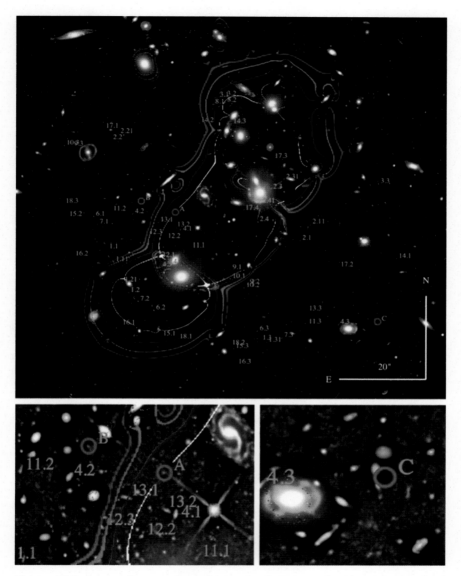

Figure 4.11 A very distant galaxy, detected only because its (three) images (red circles) are magnified by a factor of ten. Top panel shows the full field, with numbers representing galaxies used to construct the mass map; critical curves determined by multiple imaging of closer objects shown in different colors depending on the distance of the source; and the three images of the background galaxy itself indicated by the red circles (Zitrin et al., 2014). Different colored critical curves correspond to different distances to the background galaxy. Bottom panels zoom in to the three images; the bottom-left panel shows the two nearby images, which should be nearly equidistant from the critical curves. This rules out the blue and white curves that correspond to background redshifts of 1.3 and 2.

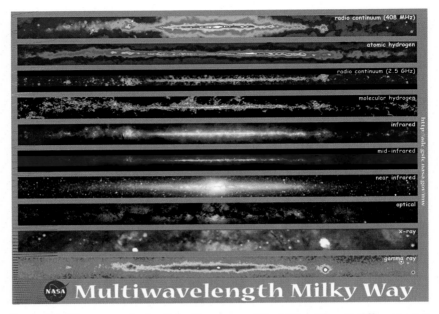

Figure 5.3 View towards the Galactic center of the Milky Way at different wavelengths. The near-infrared radiation comes predominantly from giant stars and shows a bulge that tapers off at ~ 1 kpc. Emission from hydrogen, both molecular and atomic, traces the gas and extends further, out to about 5 kpc. There is a clear fall-off at larger radii, indicating that most of the emitters are confined to a disk with a radius of about 5 kpc.

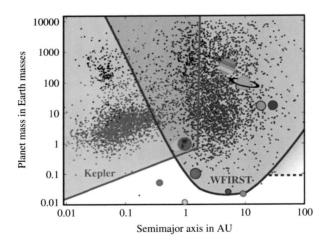

Figure 5.8 Mass and orbital radius of exoplanets discovered by the different techniques (Spergel et al., 2015). The Solar System planets are depicted pictorially. Kepler uses the transit technique and detected planets depicted by the red dots. WFIRST is a proposed satellite mission that will employ microlensing to extend the sensitivity to Earth-mass planets, as indicated by the blue shaded region; blue points are simulated detections with WFIRST. Black points are planets detected using only radial velocities.

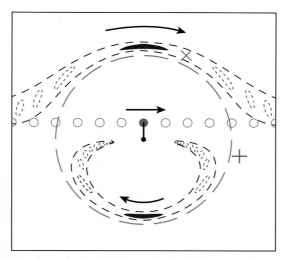

Figure 5.9 Evolution of a planetary microlensing event (Gaudi, 2010). Foreground star is black dot in the center, and open circles denote successive positions of a background star as it moves across the line of sight. The long dashed circle is the Einstein ring of the lens. There are two images of the background star, one outside the Einstein radius (top dotted ovals) and one inside. If the planet lies close to one of these images, then the background star will be further magnified for the brief duration of the overlap. A planet at the position of the cross therefore leads to an observed excess magnification while one at the position of the plus sign does not.

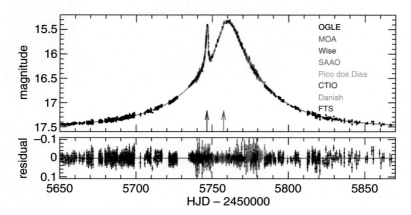

Figure 5.11 Light curve of a star in the Galactic bulge that is magnified over the course of a few months by an intervening star and planet (Skowron et al., 2015). The initial event was detected by the OGLE collaboration but then followed up with multiple observations at several different telescopes listed in the figure. The blip at Heliocentric Julian Date (HJD) $\simeq$ 5746 days is due to the Jupiter-sized planet. The four-day blip induced by the planet is well fit by the solid curve, as seen by the difference between the data and the model in the bottom panel.

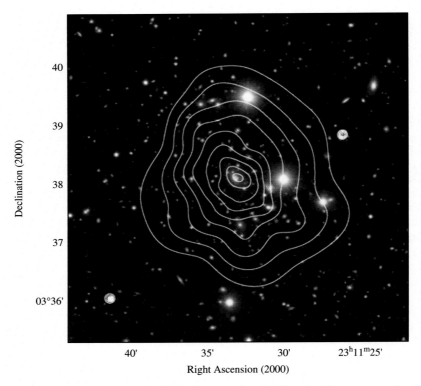

Figure 7.7 Emission from A2552 in X-ray and optical bands. The contours depict increasing X-ray flux towards the cluster center as measured by the Chandra satellite, while the colored image is based on optical observations with the University of Hawaii 2.2 meter telescope. The optical image reveals many individual galaxies in the cluster (which is at redshift $z = 0.3$), while the X-ray contours trace the hot gas that is concentrated in the cluster center. From Ebeling et al. (2010).

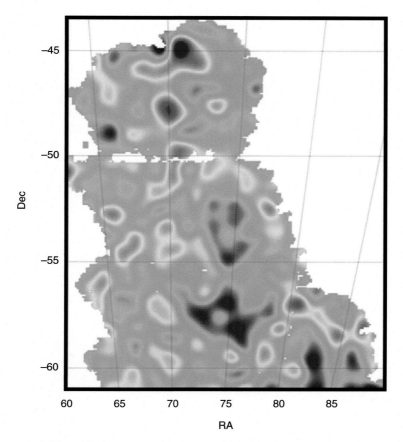

Figure 8.6 Map of convergence κ over 150 square degrees of early data from the Dark Energy Survey (Vikram et al., 2015) constructed from shapes of over a million background galaxies. Blue regions are underdense and red overdense, with the extreme values of κ reaching to ± 0.015. This map includes contributions from all matter along the line of sight to the background galaxies, which are distributed around $z \simeq 0.8$.

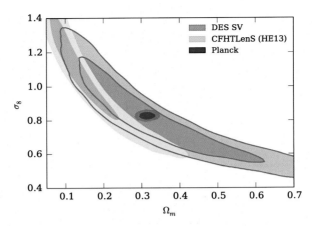

Figure 8.16 Constraints on cosmological parameters from two tomographic lensing surveys DES and CFHT (Abbott et al., 2016). The "SV" label on the DES results indicate that they come from Science Verification data, only 3 percent of the expected total data set; "HE13" refers to the paper that analyzed the CFHT data (Heymans et al., 2013). As of 2015, lensing constraints were still not as powerful as those from the cosmic microwave background, as indicated by the "Planck" allowed region.

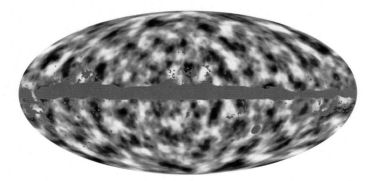

Figure 9.4 A map of the projected gravitational potential inferred from a quadratic estimator of the CMB temperature by the Planck satellite. Copyright: ESA and the Planck Collaboration.

Let us compute the changes in the ellipticity due to the changes in the moments caused by lensing. As an example, we will focus on ϵ_1. Then from the definition, the change in ϵ_1 due to lensing is

$$
\begin{aligned}
\delta\epsilon_1 &= \frac{\delta Q_{11} - \delta Q_{22}}{T} - \frac{\delta T}{T^2}(Q_{11} - Q_{22}) \\
&= \frac{\Psi_{lm} Z_{lm11} - \Psi_{lm} Z_{lm22}}{T} - \epsilon_1 \Psi_{lm} \frac{[Z_{lm11} + Z_{lm22}]}{T}.
\end{aligned} \tag{6.25}
$$

Since we kept only the first-order term in the Taylor expansion in Eq. (6.21), the change in the ellipticity is linear in the components of the distortion tensor so in principle could depend on γ_1, γ_2, and κ. In this linear limit, though, the dependence on κ is very weak (Exercise (6.7)) so we can approximate:

$$
\delta\epsilon_1 = P_{1i}\gamma_i \tag{6.26}
$$

where the 2×2 matrix P_{ij} is called the *shear polarizability*.

We will presently turn to calculate the shear polarizability in terms of moments of the surface brightness weighted by the weighting function, but first let's pause to applaud the inspiration of those who introduced this terminology (Kaiser et al., 1995). The word *polarizability* is often used in physics when describing how atoms react to an electric field. *Electric polarizability* is the proportionality constant between the observed dipole moment of an atom and the external electric field that is distorting its shape. How wonderful, then, that this same terminology is used in weak lensing: the observable in this case is the ellipticity of the galaxy, while the external field is the shear field. The shear polarizability is then the proportionality constant, just as in the electric case.

One part of the shear polarizability is proportional to the intrinsic ellipticity ϵ_1 (the second term in Eq. (6.25)) while the other is independent of ϵ_1. So, generally we can decompose the shear polarizability into two pieces:

$$
P_{ij} = X_{ij} - \epsilon_j^\gamma \epsilon_i. \tag{6.27}
$$

In the case we are focusing on, ϵ_1, the relevant components are X_{11} (the coefficient of γ_1); X_{12} (the way in which ϵ_1 depends on γ_2), and $\epsilon_{1,2}^\gamma$ (the term that conveys the crosstalk between intrinsic ellipticity and shear).

Picking out the coefficients of γ_1 in the first term in Eq. (6.25) means setting $\Psi_{lm} \to \delta_{lm}[\delta_{l1} - \delta_{l2}]$, so the numerator becomes $Z_{1111} - Z_{2211} - Z_{1122} + Z_{2222}$. Let's work out the terms that have W (as opposed to W') in the integral. Those will be

$$
[Z_{1111} - Z_{2211} - Z_{1122} + Z_{2222}]\big|_W = \int d^2\theta \mathcal{S}(\vec\theta)W(\theta)\left[3\theta_1^2 - \theta_1^2 - \theta_2^2 + 3\theta_2^2\right]
$$
$$
= 2T. \tag{6.28}
$$

So this term, when divided by T, simply gives the factor of 2 we obtained in the ideal case. Going through similar steps for the W' term leads to

$$X_{11} = 2 + 2\frac{\int d^2\theta\, S(\vec{\theta})\, W'(\theta)\, (\theta_1^2 - \theta_2^2)^2}{\int d^2\theta\, S(\vec{\theta})\, W(\theta)\theta^2}. \tag{6.29}$$

Similarly, the off-diagonal term relating ϵ_1 to γ_2 is

$$X_{12} = 4\frac{\int d^2\theta\, S(\vec{\theta})\, W'(\theta)\theta_1\theta_2\, (\theta_1^2 - \theta_2^2)}{\int d^2\theta\, S(\vec{\theta})\, W(\theta)\theta^2}. \tag{6.30}$$

Finally, the term that couples shear and ellipticity has coefficients

$$\epsilon_1^\gamma = 2\epsilon_1 + 2\frac{\int d^2\theta\, S(\vec{\theta})\, W'(\theta)\theta^2\, (\theta_1^2 - \theta_2^2)}{\int d^2\theta\, S(\vec{\theta})\, W(\theta)\theta^2}$$

$$\epsilon_2^\gamma = 2\epsilon_2 + 4\frac{\int d^2\theta\, S(\vec{\theta})\, W'(\theta)\theta^2\, \theta_1\theta_2}{\int d^2\theta\, S(\vec{\theta})\, W(\theta)\theta^2}. \tag{6.31}$$

The first term in Eq. (6.29) recaptures our initial result, that in the simple case of no weighting and a circular galaxy, the ellipticity is equal to twice the shear. All the other terms – the second one in Eq. (6.29) and X_{12}, ϵ_1^γ, ϵ_2^γ – account for the complication of intrinsic ellipticity, as they all vanish if the intrinsic galaxy shape is circular. One point to note – one that remains true for most algorithms designed to extract the shear from the measured ellipticity – is that the coefficients depend on higher moments. The two components we are after, ϵ_1 and ϵ_2, are second moments in that the integrals that define them are quadratic in θ. We see that the effect that shear has on ellipticity depends on integrals that are quartic in θ.

The shear polarizability tensor can in principle be estimated directly from the data, by computing the integrals here, or their generalizations as captured in Eq. (6.57), for each galaxy. Then, the relation

$$\epsilon_i = \epsilon_i^{(0)} + P_{ij}\gamma_j \tag{6.32}$$

could be inverted to obtain an estimate for the shear: $\hat{\gamma}_i = P^{-1}{}_{ij}e_j$ since the mean of the intrinsic ellipticity $\epsilon^{(0)}$ vanishes. In practice, this turns out to be too noisy, so other methods must be used to obtain a shear estimate. One approach is to take a very high-resolution image (from space, for example) and then degrade it, thereby calibrating the relationship. Another is to approximate the second term in Eq. (6.27) as diagonal with a value given by $-2\langle\epsilon_1^2\rangle = -2\langle\epsilon_2^2\rangle \equiv -2\epsilon_{\text{RMS}}^2$. Then the shear estimate becomes

$$\hat{\gamma}_i \simeq \frac{\epsilon_i}{\mathcal{R}} \tag{6.33}$$

where the *responsivity* $\mathcal{R}$ is defined as

$$\mathcal{R} \equiv 2\left[1 - \epsilon_{\text{RMS}}^2\right]. \tag{6.34}$$

6.5 Correcting for the Point Spread Function

Any optical system produces a slightly distorted image of its source. These distortions are due to aberrations and diffraction. When the system is ground-based, light from the source has to travel through the atmosphere and is deflected and distorted further. All of these effects are encoded in the *point spread function* (PSF), which can be described by a function $\mathcal{P}(\vec{\theta})$. The observed image is the true image in the sky convolved with the point spread function. To correct for the effects of this convolution and recapture the true image, one would like to look for objects known to be point sources. Then, their images would trace out the function $\mathcal{P}(\vec{\theta})$. This is one of the rare times in astronomy where the ideal is possible, for stars are point sources and there are many stars available from which to measure the PSF. Fig. 6.4 shows an example: the left-hand panel depicts the effect of the PSF on a star; the middle panel is an elliptical galaxy before the PSF has been applied; and the rightmost panel shows that the PSF does indeed distort the galaxy's image. Were we to apply the techniques of the previous section without correcting for the PSF, we would derive an incorrect value for the galaxy's ellipticity. The purpose of this section is to spell out one way of correcting for the effects of the PSF on the shapes of background objects.

The PSF distorts the observed shape of images so that the observed surface brightness is

$$S^{\text{obs}}(\vec{\theta}) = \int d^2\theta' \mathcal{P}(\vec{\theta}')S(\vec{\theta} - \vec{\theta}'). \tag{6.35}$$

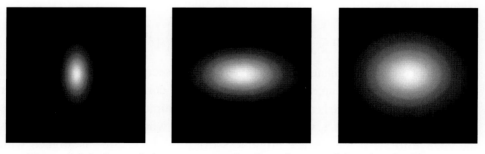

Figure 6.4 Left panel shows an example of a point spread function: in this case, a point-like star appears extended and elliptical. The right panel depicts the effect of this PSF on the galaxy depicted in the middle panel. The ellipticity of the PSF acts against the galaxy's ellipticity (in this case) and puffs up the image, so would distort our extraction of shear unless corrected.

We can Taylor expand the surface brightness around $\vec{\theta}$. The first term gives the undistorted surface brightness as we normalize $\mathcal{P}$ so that it integrates to unity: $\int d^2\theta \mathcal{P}(\theta) = 1$. We can also choose the coordinate system so the first moment vanishes: $\int d^2\theta \mathcal{P}(\theta)\vec{\theta} = 0$. This leaves the second-order term in the Taylor expansion as the leading correction to the observed surface brightness.

$$\mathcal{S}^{\text{obs}}(\vec{\theta}) \simeq \mathcal{S}(\vec{\theta}) + \frac{1}{2}\frac{\partial^2 \mathcal{S}}{\partial\theta_l \partial\theta_m}\bigg|_{\vec{\theta}} \int d^2\theta' \mathcal{P}(\vec{\theta}')\theta_l' \theta_m'$$

$$\equiv \mathcal{S}(\vec{\theta}) + \frac{1}{2}\frac{\partial^2 \mathcal{S}}{\partial\theta_l \partial\theta_m}\mathcal{P}_{lm} \tag{6.36}$$

where the $\mathcal{P}_{lm}$s are the second moments of the PSF. These moments characterize the shape of the PSF and therefore govern the distortions in the observed surface brightness patterns. What we want to do is measure these characteristics via the shapes of observed stars and then subtract off the distortions.

As in Eq. (6.17), we form the weighted moments of the observed surface brightness, but now accounting for the distortion due to the PSF:

$$Q'_{ij} = Q_{ij} + \frac{\mathcal{P}_{lm}}{2}\int d^2\theta\, W(\theta)\theta_i \theta_j \frac{\partial^2 \mathcal{S}}{\partial\theta_l \partial\theta_m} \tag{6.37}$$

so the PSF distortion is encoded in the second term. Integrating by parts twice leads to

$$Q'_{ij} = Q_{ij} + Y_{lmij}\mathcal{P}_{lm} \tag{6.38}$$

with

$$Y_{lmij} \equiv \frac{1}{2}\int d^2\theta\, \mathcal{S}(\vec{\theta})\frac{\partial^2}{\partial\theta_l \partial\theta_m}\left[W(\vec{\theta})\,\theta_i \theta_j\right]. \tag{6.39}$$

Let's first consider the effects of a circular PSF, so that $\mathcal{P}_{lm} = \delta_{lm}\sigma^2_{\text{PSF}}$. Then, the change in the numerator of ϵ_1 for example is equal to

$$\Delta\left(Q_{11} - Q_{22}\right) = \sigma^2_{\text{PSF}}\frac{1}{2}\int d^2\theta\, \mathcal{S}(\vec{\theta})\left\{\frac{\partial^2}{\partial\theta_1^2} + \frac{\partial^2}{\partial\theta_2^2}\right\}\left[W(\vec{\theta})\,(\theta_1^2 - \theta_2^2)\right]. \tag{6.40}$$

This will of course propagate to the change in ϵ_1. The second derivatives in brackets comprise the 2D Laplacian, which measures the local curvature of a function. Near a maximum the curvature is negative, while near a minimum it is positive. With this in mind, consider Fig. 6.5, which plots the function whose Laplacian determines the change to ϵ_1. If the surface brightness from a galaxy, $\mathcal{S}(\vec{\theta})$ is aligned with the θ_1 axis, then the integral in Eq. (6.40) will pick out the two positive peaks in Fig. 6.5. Since both of these peaks have negative curvature, the change in ϵ_1 due to the PSF will be negative. That is, in the case when $\epsilon_1 > 0$ (galaxy elongated along the θ_1 axis), a circular PSF will act to *reduce* the ellipticity. Similarly, if the

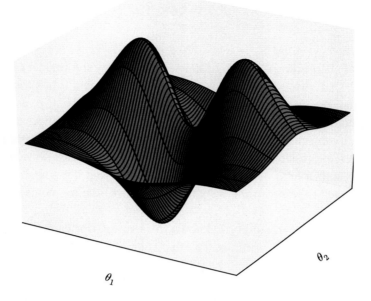

θ_2

θ_1

Figure 6.5 The function $W(\vec{\theta})[\theta_1^2 - \theta_2^2]$ that is differentiated in Eq. (6.40) to obtain the impact of the PSF on the ellipticity ϵ_1. Here a Gaussian window function W has been assumed.

galaxy were elongated along the θ_2 axis so that ϵ_1 was negative, then the integral in Eq. (6.40) would pick out the negative peaks in Fig. 6.5: those with positive curvature. So, again, the magnitude of ϵ_1 would be reduced. This pictorial argument is quite general: a circular PSF acts to dilute the observed ellipticity. We can even guess at the magnitude of the effect: the derivatives in Eq. (6.40) will act to bring down two factors of the width of the window function, which is usually chosen to be roughly the angular size of the galaxy, θ_{gal}. Therefore, the ellipticity will be diluted by a factor of order $\sigma_{PSF}^2/\theta_{gal}^2$. Exercise (6.10) works through a simple example more quantitatively, but since the PSF is often roughly the same size as the galaxies whose shapes are being measured, clearly quite a bit of care must be taken to account for the PSF.

As depicted in Fig. 6.4, when the PSF is elliptical, the impact on the inferred shear is even more dangerous. Very similar to the shear polarizability, we can define the *smear polarizability* such that the change in ellipticity due to the PSF is

$$\Delta\epsilon_i = P_{ij}^s p_j. \qquad (6.41)$$

This equation captures the way that the PSF affects the ellipticity just as Eq. (6.26) captured the way that the ellipticity responds to an external shear. The role of γ_i here is played by the appropriate combinations of the moments of the PSF:

$$p_1 = \mathcal{P}_{11} - \mathcal{P}_{22}$$
$$p_2 = 2\mathcal{P}_{12}. \tag{6.42}$$

So, Eq. (6.41) is a powerful yet familiar way to convey the impact of the PSF on ellipticities. That virtue mitigates the annoying feature that it has too many "p"s. To rehash their defintions: $\mathcal{P}_{ij}$ are the moments of the PSF $\mathcal{P}(\vec{\theta})$; p_i as defined here are simply convenient combinations of $\mathcal{P}_{ij}$; and P^s_{ij} is the smear polarizability, the proportionality constant that relates ellipticity to the PSF. The algebra required to obtain explicit expressions for P^s_{ij} is identical to that in the last section, so we will not walk through it here. It should be clear, though, that the smear polarizability can be computed in terms of the weighted integrals of the moments of the surface brightness.

Once the smear polarizability is known, it is straightforward to correct galaxy ellipticities for the PSF distortions. The first step is to use the measured shapes of the stars to infer the properties of the PSF encoded in the p_i. Consider the left panel in Fig. 6.6, which comes from Hoekstra et al. (1998). The stars should appear as point sources, but instead show up as elliptical objects with ellipticities exceeding 10 percent; they are tracing the PSF and therefore the p_i. Note that these moments vary with position in the field. At the top and bottom of the chips, the PSF $p_1 \simeq 0.1$ and $p_2 \simeq 0$ while on the sides $p_1 \simeq -0.1$. Clearly, then, the p_i need to be measured everywhere in the region of interest. This motivates the following recipe:

- Infer the PSF moments p_i from the ellipticities of stars, ϵ^*_i, in the field. The method proposed in Kaiser et al. (1995) was to take the stellar smear polarizability to be diagonal so that

$$p_i \simeq \frac{\epsilon^*_i}{P^{s,*}_{ii}} \qquad \text{No Sum} \tag{6.43}$$

  where the superscript denotes the stars.
- Interpolate these moments across the field in a polynomial: $p_i(\vec{\theta}) = a_i + b_{i,x}\theta_x + b_{i,y}\theta_y + \ldots$. Fit the coefficients by using the measured p_i of each star.
- Correct the ellipticities of objects by subtracting the moments of the PSF:

$$\epsilon_i \to \epsilon_i - P^s_{ij} p_j(\vec{\theta}). \tag{6.44}$$

The smear polarizability of the galaxies here can be measured by carrying out the weighted integrals over the surface brightness. The right panels of Fig. 6.6 show that the corrected ellipticities are, for the most part, free of the effects of the PSF.

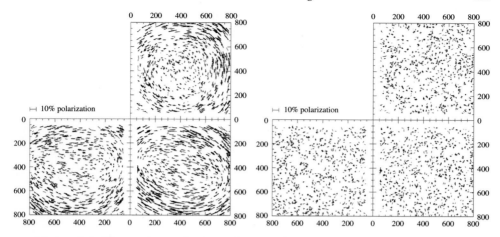

Figure 6.6 Observations of galaxy shapes using the Wide Field and Planetary Camera 2 on the Hubble Space Telescope (Hoekstra et al., 1998). Left panel shows the ellipticities of stars in three different chips. Even though stars should be seen as point sources, the point spread function distorts their images. The right panel shows the residual ellipticities after correcting for the effects of the PSF.

6.6 Forward Modeling

The method described in §6.5, which relies on moments, suffers from several problems. Consider the moments of the PSF defined in Eq. (6.42). They are unweighted, which opens up the possibility that a PSF can have vanishing moments but nonzero ellipticity as a function of position in the galaxy. Fig. 6.7 shows a simple example adapted from Kuijken (1999). This is an artificial example, but with quite realistic ramifications. In the innermost regions, the integrand of p_1, $\mathcal{P}(\vec{\theta})[\theta_x^2 - \theta_y^2]$, is positive, but in the outermost regions, it is negative. Integrated over all space, as required by Eq. (6.42), the moments (both p_1 and p_2) would vanish. So the moment method, as quantified in Eq. (6.41), would predict that a circular galaxy convolved with this PSF would produce zero ellipticity. But when convolved with a real galaxy distribution, these local distortions would produce non-vanishing ellipticity even if the galaxy were circular. You can work through this in Exercise (6.13). So, this is a failing of the formalism that corrects for the PSF by using its unweighted moments.

There are ways to fix this problem that adhere to the general idea of estimating the ellipticity by measuring moments, but historically, the realization of this problem led to a completely new way of inferring ellipticity, a technique now known as *forward modeling*. Let's walk through a simple example that has been successfully implemented, but you should be warned that the techniques are evolving rapidly, and even this method, dubbed im2shape (Bridle et al., 2002), has been updated.

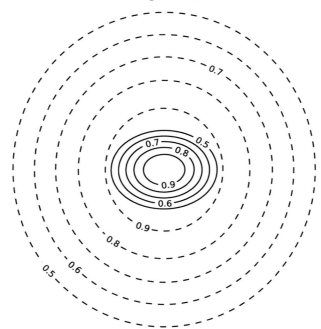

Figure 6.7 An example of an artificial PSF, which is the sum of the two elliptical Gaussians depicted by the contours. The solid contours reflect a Gaussian with a small radius, elliptically oriented along the x-axis; the dashed a larger radius with opposite ellipticity. If the total PSF were the sum of these two, p_1, as defined in Eq. (6.42), would be positive if only the innermost pixels were used but would be zero if all pixels were used.

The overarching idea is to model the galaxy surface brightness with a function that contains a set of parameters, do the same for the PSF, and then generate predictions for the observed surface brightness. These predictions will depend on the parameters used in the model, so the goal of measuring ellipticities is boiled down to estimating the parameters that best fit the data. A simple example is to model the surface brightness of the galaxy as an elliptical Gaussian:

$$S^{\text{model}}(\vec{\theta}) = \frac{A}{2\pi \, \det(C)} \exp\left\{ -\frac{1}{2} (\theta_i - \theta_i^{\text{center}}) \, C_{ij}^{-1} (\theta_j - \theta_j^{\text{center}}) \right\}. \quad (6.45)$$

This model has six parameters: the amplitude A, the two coordinates of the center of the image $\theta_x^{\text{center}}, \theta_y^{\text{center}}$, and the three components of the 2×2 covariance matrix C. To allow for all possible ellipticities and orientations, C is set to

$$C = \frac{1}{2} \left(\begin{array}{cc} a^2 \cos^2 \alpha + b^2 \sin^2 \alpha & (b^2 - a^2) \sin \alpha \cos \alpha \\ (b^2 - a^2) \sin \alpha \cos \alpha & b^2 \cos^2 \alpha + a^2 \sin^2 \alpha \end{array} \right) \quad (6.46)$$

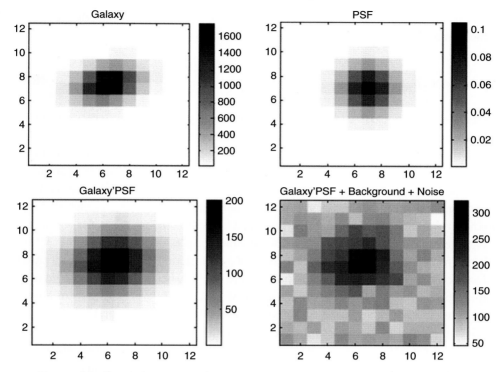

Figure 6.8 Simulation image of a galaxy assuming both galaxy and PSF are elliptical Gaussians. Upper left shows the galaxy and upper right the PSF; lower panels show the convolved image without (left) and with (right) noise. Credit: *The Shapes of Galaxies and their Dark Halos*, Bridle, S. L., Kneib, J.-P., Bardeau, S., and Gull, S. F. Editor: Natarajan, P., Copyright 2002, World Scientific.

so that the three parameters are the lengths of the long and short axes a, b and the orientation of the ellipse α. You can show how these parameters are related to the ellipticity of the galaxy in Exercise (6.11). As a simple example, when $\alpha = b = 0$, the second moment of θ_x is equal to $C_{xx} = Aa^2/2$ while the other moments are zero. So that limit corresponds to $\epsilon_1 = 1$; $\epsilon_2 = 0$.

The PSF is also modeled as an elliptical Gaussian, with the only difference being that the parameters are assumed to be known, as they have been measured from stars. The prediction, then, for the intensity at every position in the field reduces to an integral over a Gaussian galaxy model convolved with a Gaussian PSF. A Gaussian convolved with a Gaussian is another Gaussian (see Exercise (6.12)), so the resulting prediction for each pixel is a simple analytic function of the parameters. An example of this from Bridle et al. (2002) is shown in Fig. 6.8.

Given the final image that would be observed, in the lower-right panel of Fig. 6.8, a human would be hard pressed to extract the input values of the galaxy's ellipticities: $\epsilon_1 = 0.2$ and $\epsilon_2 = 0.1$. But simple model fitting does indeed extract the correct input values. The way to do this is to compute

$$\chi^2\left(a, b, \alpha, \vec{\theta}^{\text{center}}, A\right) \equiv \sum_{i=1}^{N_{\text{pixels}}} \frac{\left(\mathcal{S}(\vec{\theta}_i) - \mathcal{S}^{\text{model}}(\vec{\theta}_i; a, b, \alpha, \vec{\theta}^{\text{center}}, A)\right)^2}{\sigma^2} \quad (6.47)$$

where σ is the noise in each pixel. If there were only one free parameter, this would be simple computationally: one would just evaluate the χ^2 for, say, 20 values of the parameter, see where it is lowest and constrain the parameter by how rapidly the χ^2 increases away from its minimum. However, in this case, there are six parameters so a similar approach would require 20^6 evaluations: not impossible but certainly challenging! And this is for a simple model, where the galaxy is modeled by a single Gaussian. More generally, several Gaussians are used, so the number of parameters quickly multiplies and it becomes impossible to sample the χ^2 in a brute-force way. There are several clever ways to sample the space, taking care to focus on the region in parameter space where the fit is best. Many of these go under the rubric *Markov Chain Monte Carlo* or MCMC. They typically start at a point, move to another point based on some rule and random number, and then accept or reject the point with a probability related to how much better or worse the χ^2 is. Using an MCMC, the developers of im2shape were able to extract ellipticities of galaxies in the complex environment of ground-based observations.

6.7 GREAT

The problem of measuring shapes is so difficult that many approaches have been proposed. Members of the lensing community foresaw a need to test these different approaches, so developed the GRavitational lEnsing Accuracy Testing (GREAT) program. The basic idea is to simulate what measurements from surveys might look like and test the different codes. Since the input shears are known, there is a clear way to distinguish which methods are most effective. The program has completed three rounds of tests (Mandelbaum et al., 2014) and has become an integral part of all analyses.

The quantitative metric employed in the GREAT program and in most analyses includes the possibilities that the measured shears recover the true shears multiplied by some constant not equal to one and that the measured shears are off by a constant. That is, the measured shears are

$$\frac{\gamma_i^{\text{measured}}}{1 - \kappa} = m_i \frac{\gamma_i^{\text{true}}}{1 - \kappa} + c_i \quad (6.48)$$

with m_i and c_i called the multiplicative and additive bias, respectively. Note that it is the reduced shears that are used, although the difference between the reduced and bare shears is irrelevant for most weak lensing applications. The exact metric has varied over the GREAT challenges, but they all award the most points for methods that produce the smallest values of $m_i - 1$ and c_i.

There are two types of errors in weak lensing: statistical and systematic. Statistical errors are caused by the intrinsic ellipticities of galaxies and, as we will see in the following chapter, require observers to measure many background galaxies in order to extract the usually small signal from this source of noise. The defining feature of a statistical error is that it goes down as more measurements are made. Systematic errors are more pernicious in that they remain to contaminate results even if many measurements are made. The multiplicative and additive biases quantify systematic errors, and attempting to minimize them is the mission of the team that put together the GREAT challenges. We have touched on several potential sources of these systematic errors in this chapter: the PSF, the weighting function required by noise, and the polarizabilities. The end-to-end work of the GREAT team addresses these and others, so provides evidence that we will be able to extract the information encoded in galaxy shapes. With an understanding of how difficult it is to measure these shapes, we now turn to some potential pay-offs.

Suggested Reading

Detection of weak gravitational lensing by observing the distorted shapes of many background objects came about ten years after the first multiply imaged system was detected in 1979. For example, Tyson et al. (1990) observed the distortion of background galaxies behind galaxy clusters, a topic we will explore in the next chapter. Theorists showed (Blandford et al., 1991; Miralda-Escude, 1991) that the effect could be measured even without an obvious massive object in the foreground; i.e., weak lensing by large-scale structure could be observed.

The review paper of Bartelmann and Schneider (2001) is comprehensive and contains a very nice section (§4.6) on shape measurement, while the more recent review of Hoekstra and Jain (2008) is less technical but with a more updated overview of what is expected from upcoming surveys. For more technical papers, Kaiser et al. (1995) worked through many of the complications of measuring shapes; their method for obtaining shape estimates, followed in §6.4 and §6.5, is called the KSB method. Numerous alternatives and improvements have been proposed and implemented since, but I think learning the KSB method is a useful starting point to the literature. The notion of using an elliptical Gaussian weighting function to capture the bulk of the signal while reducing the noise was introduced in Bernstein and Jarvis (2002). First the shape is measured; then a weighting function matching the shape is chosen; then the shape is re-measured with the new elliptical weighting function; and the process iterates until it converges. This idea was further refined by Hirata and Seljak (2003), who used non-Gaussian functions for the weighting. Indeed, the choice of *basis* functions with which to weight the surface brightness is itself a mini-industry, dating back to the proposal of Refregier (2003) to use *shapelets*. A good, short, recent description of modern implementations of the moments method is given by Melchior et al. (2011).

Kuijken (1999) introduced the forward modeling technique as an alternative to taking moments, and that approach has developed as perhaps the preferred alternative, as evidenced, for example, by the first analysis of shapes in the Dark Energy Survey (Jarvis et al., 2015), in which two techniques were used, both of which involved forward modeling. There are many resources explaining the idea of a Markov Chain Monte Carlo. To pick one clear example, Foreman-Mackey et al. (2013) explain the method and provide an implementation that is frequently updated and freely available.

The GREAT3 challenge is described in Mandelbaum et al. (2014), which – in §2.1 – also does a very nice job of accumulating and explaining the different conventions used to describe shear and ellipticity. More generally, all of the literature produced in the GREAT challenges is a pleasure to read, filled with informative figures, and all the information one needs to get started in this very challenging field.

Exercises

6.1 Show that the distortion in the shape of a circular object due to non-zero γ_2 is given by Eq. (6.5).

6.2 Imagine rotating the x–y axes by an angle θ. The goal of this exercise is to understand how the elements of the distortion tensor change. First, recall that the rotation matrix

$$U(\theta) \equiv \begin{pmatrix} \cos\theta & -\sin\theta \\ \sin\theta & \cos\theta \end{pmatrix} \tag{6.49}$$

transforms a vector $(x, 0)$, say into one with coordinates $(x\cos\theta, x\sin\theta)$. So if θ were equal to $\pi/2$, for example, a vector previously aligned with the x-axis would now be aligned with the y-axis. That is, this rotation matrix does indeed describe a rotation of the coordinate system by an angle θ about the x-axis. What happens to the distortion matrix under this rotation? Compute

$$\Psi' = U^T \Psi U \tag{6.50}$$

with components $\Psi'_{ij} = U_{ik}\Psi_{kl}U_{lj}$. What are the new values of κ', γ'_1, and γ'_2 under this rotation? Show that the complex shear

$$\gamma \equiv \gamma_1 + i\gamma_2 \tag{6.51}$$

transforms as a spin-2 field; i.e, upon rotation by an angle θ it transforms as $\gamma \rightarrow e^{2i\theta}\gamma$.

6.3 The general expression for the relation between shear and ellipticity is given by

$$\epsilon^{(0)} = \frac{\epsilon - 2g + g^2\epsilon^*}{1 + |g|^2 - 2\text{Re}\left[g\epsilon^*\right]}, \tag{6.52}$$

where $\epsilon^{(0)}$ is the galaxy's ellipticity in the absence of lensing; ϵ after lensing; and the reduced shear is defined as

$$g \equiv \frac{\gamma}{1 - \kappa}. \tag{6.53}$$

In all of these expressions, g, ϵ, γ, and $\epsilon^{(0)}$ are complex quantities (hence the * denotes complex conjugate) defined via $\gamma = \gamma_1 + i\gamma_2$, for example. Show that this expression is consistent with our result in the simple case when $\epsilon^{(0)} = 0$, the background galaxy is intrinsically circular.

6.4 Suppose a galaxy has a half-light radius of 10 kpc. What angle does it subtend if it is 100 Mpc from us? 1 Gpc?

6.5 Starting from the definition in Eq. (6.7), compute ϵ_1 in terms of the shear and convergence. Show that it is equal to the expression in Eq. (6.15).

6.6 Replace the denominator in Eq. (6.7) with

$$q_{11} + q_{22} \rightarrow q_{11} + q_{22} + 2\sqrt{q_{11}q_{22} - q_{12}^2} \tag{6.54}$$

and show that the relation between ellipticity and shear (for a circular background galaxy) no longer has the factor of 2 that appears in front of Eqs. (6.14) and (6.15).

6.7 We derived an expression for the change in ϵ_1 in terms of the distortion tensor:

$$\delta\epsilon_1 = \frac{\Psi_{lm}Z_{lm11} - \Psi_{lm}Z_{lm22}}{T} - \epsilon_1\Psi_{lm}\frac{[Z_{lm11} + Z_{lm22}]}{T}. \tag{6.55}$$

Show that the right-hand side here depends only very weakly on the convergence κ. Start by setting above $\Psi_{11} = \Psi_{22} = \kappa$ with the off-diagonal elements equal to zero. (We have already computed the dependence of $\delta\epsilon_1$ on shear, so can you ignore those here.) Show that the sum of terms in the integrand weighted by $\mathcal{S}W$ vanish, while those weighted by $\mathcal{S}W'$ are proportional to the difference

$$\langle(\theta_1^2 - \theta_2^2)(\theta_1^2 + \theta_2^2)\rangle - \epsilon_1\langle(\theta_1^2 + \theta_2^2)^2\rangle \tag{6.56}$$

where $\langle\ldots\rangle = \int d^2\theta\,\mathcal{S}(\vec{\theta})W'(\theta)\ldots$. Argue that the difference above is very close to zero.

6.8   Show that the full expression for the 2×2 matrix that relates ellipticity to shear is

$$X_{ij} = \frac{2}{T} \int d^2\theta \, \mathcal{S}(\vec{\theta}) \begin{bmatrix} W\theta^2 + W'(\theta_1^2 - \theta_2^2)^2 & 2W'\theta_1\theta_2(\theta_1^2 - \theta_2^2) \\ 2W'\theta_1\theta_2(\theta_1^2 - \theta_2^2) & W\theta^2 + 4W'\theta_1^2\theta_2^2 \end{bmatrix}.$$

(6.57)

6.9   The claim is that the PSF can be inferred by measuring the shapes and sizes of stars. This relies on stars being effectively point-like objects on the sky. Convince yourself of this by computing the angular size of a star like the Sun that is a distance 100 pc from us. Show that this angular size is much smaller than an arcsecond, which is roughly the size of the PSF in many surveys.

6.10 Consider the change to $Q_{11} - Q_{22}$ due to a circular PSF as given in Eq. (6.40) in the special case when the galaxy is completely aligned along the θ_1-axis, so that $\epsilon_1 = 1$. That is, choose

$$\mathcal{S}(\vec{\theta}) = \frac{I_0}{\theta_{\text{gal}}} \, \delta_D(\theta_2) \, e^{-\theta_1^2/2\theta_{\text{gal}}^2}$$

(6.58)

where $\delta_D(\theta_2)$ is a Dirac delta function that confines the surface brightness to the θ_1-axis. Take $W(\vec{\theta})$ to be a symmetric Gaussian with width θ_{gal}. Don't worry about the normalization since it will drop out of the ratio, so simply set $W = e^{-\theta^2/2\theta_{\text{gal}}^2}$. First determine $Q_{11} - Q_{22}$ in the absence of the PSF, and show that it is equal to

$$Q_{11} - Q_{22} = \frac{\pi^{1/2}\theta_{\text{gal}}^2 I_0}{2}.$$

(6.59)

Of course, this is equal to $Q_{11} + Q_{22}$, so, in the absence of the PSF, $\epsilon_1 = 1$. Now compute the change given by Eq. (6.40) and show that

$$\frac{\Delta(Q_{11} - Q_{22})}{Q_{11} - Q_{22}} = -\frac{9\sigma_{\text{PSF}}^2}{4\theta_{\text{gal}}^2}.$$

(6.60)

Finally, include the change in T to complete the proof that ϵ_1 decreases.

6.11 If the surface brightness of a galaxy is described by Eq. (6.45) with $\vec{\theta}^{\text{center}} = 0$, calculate the ellipticities ϵ_1 and ϵ_2 in terms of the parameters a, b, α.

6.12 Show that a Gaussian convolved with a Gaussian yields another Gaussian. I.e., calculate

$$f(x) = \int_{-\infty}^{\infty} dx' \, e^{-(x-x')^2/2\sigma_1^2} \, e^{-(x')^2/2\sigma_2^2}$$

(6.61)

and show that $f(x)$ is a Gaussian function.

6.13 Consider a PSF that is the sum of two elliptical Gaussians:

$$P(\vec{\theta}) = \frac{1}{2\pi\sigma^2(1+\delta)} \exp\left\{-\frac{\theta_x^2}{2\sigma^2(1+\delta)^2} - \frac{\theta_y^2}{2\sigma^2}\right\}$$

$$+ \frac{1}{2\pi\sigma^2(4-\delta)} \exp\left\{-\frac{\theta_x^2}{2([4-\delta]\sigma)^2} - \frac{\theta_y^2}{2(4\sigma)^2}\right\}. \quad (6.62)$$

This is the PSF depicted in Fig. 6.7 with $\delta = 0.3$. Show that the moments, p_1 and p_2, of this PSF vanish when the integral is over all space.

Now convolve a galaxy with surface brightness $S(\vec{\theta}) = I_0 e^{-\theta^2/2\sigma^2}$ with this PSF and show that, even though the galaxy is intrinsically circular and the moments of the PSF are zero, the observed ellipticity will be nonzero.

7

Mass from Shapes

With the nitty-gritty work of measuring galaxy shapes and inferring the shear from them behind us, it is time to tackle the question of what can be learned from the shear field. The next three chapters address this in increasing generality. Here, we focus on the case when a single lens is responsible for the shear. Although never strictly speaking true, a single very massive object can indeed provide the dominant contribution to the deflection of light from the galaxies behind it. This is especially relevant for the largest bound objects in the universe, galaxy clusters.

We will see that weak lensing offers a powerful way of inferring the mass of the lens, so measuring the shapes of background galaxies has emerged as one of the leading ways of inferring the masses of galaxy clusters. Even less massive objects like single galaxies can produce observable distortions. The key in both of these cases, though, is to average over many background objects. The signal-to-noise from a single background object is very low; indeed, this is the hallmark of weak lensing. But, by averaging over N background objects, the noise is reduced by a factor of $\sqrt{N}$ and the signal emerges. In the case of the much lighter galaxies as lenses, the foreground lenses themselves must be averaged over as well in order to extract the even smaller signal. So we are entering the regime where the pictures are less impressive but the payoff is that the number of things available to study is much larger. Think of the MACHO searches described in Chapter 5: astronomers monitored millions of objects in order to find one that is micro-lensed. Here too, to extract information about the mass of an object, we will have to measure the shapes of many background galaxies. Weak lensing shifts the focus from the one spectacularly distorted galaxy to the thousands or even millions (or more!) of only very slightly distorted background galaxies.

7.1 Shear from a Spherical Distribution

In Chapter 3, culminating in Eq. (3.19), we derived the result that the deflection angle from a spherically symmetric distribution is equal to

$$\vec{\alpha}(\vec{\theta}) = \frac{\vec{\theta}}{\theta^2} \frac{M(\theta)}{\pi D_L^2 \Sigma_{cr}} \tag{7.1}$$

where $M(\theta)$ is the mass enclosed within a cylinder of angular radius θ.

To derive the two components of shear from this deflection angle, we take the relevant combination of derivatives. First,

$$\gamma_1 = \frac{1}{2} \left(\frac{\partial \alpha_1}{\partial \theta_1} - \frac{\partial \alpha_2}{\partial \theta_2} \right). \tag{7.2}$$

The derivative acting on $\vec{\theta}$ simply leaves the remaining coefficient $M/(\pi D_L^2 \theta^2 \Sigma_{cr})$, so that this term cancels in the difference that forms γ_1. What's left then is

$$\gamma_1 = \frac{1}{2} \left(\theta_1 \frac{\partial}{\partial \theta_1} - \theta_2 \frac{\partial}{\partial \theta_2} \right) \left[\frac{M(\theta)}{\pi D_L^2 \theta^2 \Sigma_{cr}} \right]. \tag{7.3}$$

Taking the derivatives leads to

$$\gamma_1 = \frac{\theta \cos(2\phi)}{2} \frac{\partial}{\partial \theta} \left[\frac{M(\theta)}{\pi D_L^2 \Sigma_{cr} \theta^2} \right], \tag{7.4}$$

where ϕ is the angle between the x-axis and $\vec{\theta}$: $\cos \phi = \theta_x/\theta$. Similarly, the second component of shear is

$$\gamma_2 = \frac{\theta \sin(2\phi)}{2} \frac{\partial}{\partial \theta} \left[\frac{M(\theta)}{\pi D_L^2 \Sigma_{cr} \theta^2} \right]. \tag{7.5}$$

The dependence on the polar angle ϕ leads to a characteristic pattern. To get a sense of this pattern, let's consider the case where $M(\theta)/\theta^2$ decreases, a likely scenario for an overdense region. Then, for example, when $\phi = 0$ so that the background galaxy lies on the positive x-axis, γ_2 vanishes and γ_1 is negative so the shear points vertically. By contrast, when $\phi = \pi/2$, γ_2 is still 0, but (since $\cos \pi = -1$) γ_1 is now positive, so the shear is horizontal. Fig. 7.1 fleshes out the ensuing pattern of shears if the lens is a point mass.

Mathematically, the linear combination

$$\gamma_t \equiv -\gamma_1 \cos(2\phi) - \gamma_2 \sin(2\phi) \tag{7.6}$$

is independent of polar angle ϕ and quantifies the amplitude of the distortion; this is the definition of the *tangential shear*. Everywhere in an annulus with radius θ, the tangential shear produced by a spherical mass distribution is equal to

$$\gamma_t(\theta) = -\frac{\theta}{2\pi D_L^2 \Sigma_{cr}} \frac{\partial}{\partial \theta} \left[\frac{M(\theta)}{\theta^2} \right]. \tag{7.7}$$

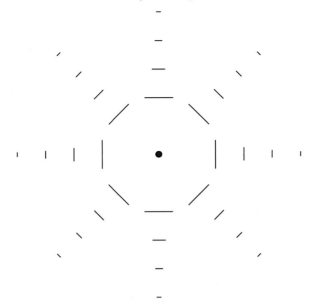

Figure 7.1 The shear pattern around a point mass. All shears are oriented perpendicular to the line pointing back to the center: the shear is *tangential*. The size of the line at any given point is proportional to the amplitude of the shear. In this case of a point mass, the shear falls off as the square of the angular distance from the lens.

Carrying out the derivative and using the definitions in Eqs. (3.18) and (3.21) leads to

$$\gamma_t(\theta) = \bar{\kappa}(\theta) - \kappa(\theta), \tag{7.8}$$

where κ is the surface density divided by the critical surface density and $\bar{\kappa}(\theta)$ is the mean value of κ within the angular radius θ. A slightly more physical quantity that is sometimes plotted is obtained by multiplying through by the critical surface density:

$$\gamma_t(R)\,\Sigma_{\mathrm{cr}} = \bar{\Sigma}(R = D_L\theta) - \Sigma(R = D_L\theta) \tag{7.9}$$

where the mean surface density $\bar{\Sigma} \equiv \Sigma_{\mathrm{cr}}\,\bar{\kappa}$.

The tangential shear is one linear combination of the two components of shear. It is useful to define another

$$\gamma_\times \equiv \gamma_1 \sin(2\phi) - \gamma_2 \cos(2\phi). \tag{7.10}$$

From Eq. (7.4) and Eq. (7.5), we see that this linear combination vanishes if the mass distribution is spherically symmetric. Therefore, $\gamma_\times$ is a useful *null test*, a diagnostic that should vanish. Because of noise, a topic we will come to shortly, $\gamma_\times$ will never be exactly zero even in the symmetric case. But, accounting for the error bars, it should be consistent with zero. This is one good reason to under-stand it. Also, we will see a more sophisticated version of $\gamma_\times$ when considering the cosmic shear pattern in the next chapter, so it is good to develop a visual sense of what these modes look like. Fig. 7.1 shows the tangential shear pattern due to an over-density; an under-density would simply have all sticks rotated by 90°. Fig. 7.2 shows the cross shear pattern that should *not* be produced by either an over- or under-density. The curl-nature of this pattern is evident; something that cannot be produced by a scalar field such as κ. Indeed, taking a cue from electromagnetism, where the electric field is the gradient of a scalar field and the magnetic field the curl of a vector field, this cross-pattern is sometimes called a "B-mode." Note that the pattern in Fig. 7.2 corresponds to rotating the shears in Fig. 7.1 by 45°.

7.2 Signal-to-Noise

The previous chapter established a path towards estimating the shear from the shapes of background galaxies. The last section established the signal in terms of the mass distribution of the lens. The final step towards analyzing lensing data is to understand the noise. In the case of lensing, the dominant form of noise is called

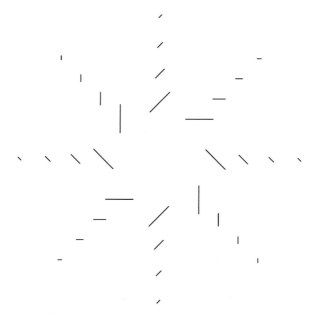

Figure 7.2 Cross shear pattern that cannot be produced by an over- or under-density of matter. If this pattern is detected at levels above the noise, there is an indication of some systematic problem afflicting the data.

shape noise due to intrinsic ellipticity as introduced in §6.4. Recall that intrinsic ellipticity is caused by the fact that, even if unlensed, galaxies would appear to be elliptical to some degree. Indeed, even if all galaxies were disks, and so appeared circular when viewed face-on – and therefore had zero ellipticity – their random orientations in the sky mean that we would observe some of them edge-on and therefore they would appear very elliptical. And, of course, not all galaxies are disk-like; some would appear elliptical even face-on. That is the bad news. The good news is that these ellipticities are random, so that the mean contribution of shape noise vanishes:

$$\langle \gamma_{\text{sh}} \rangle = 0. \tag{7.11}$$

The angular brackets here denote an average over all galaxies observed. A histogram of the ellipticities of even unlensed galaxies therefore would have mean zero, but finite width. It is this scatter, or noise, that pollutes the measurements and makes it difficult to obtain an estimate of the shear from galaxy ellipticities. Let us define the variance as

$$\sigma_{\text{sh}}^2 \equiv \langle \gamma_{\text{sh}}^2 \rangle. \tag{7.12}$$

Typical values of σ_{sh} are of order 0.2. Fig. 7.3 shows a histogram of ellipticities from one survey. The mean in the ellipticity is, as expected, equal to zero and the width is of order 0.45. Remembering the factor of 2 between shear and ellipticity, we see that the shape noise is at most 0.22, and in fact some of the width in Fig. 7.3 comes from measurement error, so 0.2 is indeed a good approximate number to have in mind for shape noise.

We can now begin to compute the signal-to-noise for a given lens. Observing one background galaxy leads to a signal-to-noise squared of $\gamma_t^2 / \sigma_{\text{sh}}^2$. But if N_a source galaxy shapes are measured, and they all share the same signal – if, for example, they all lie in an annulus a with angular radius θ_a – then the noise in that annulus will be $\sigma_{\text{sh}}/\sqrt{N_a}$ (see Exercise (4.8) for a hands-on demonstration that the noise is reduced by the square root of the number of measurements). The total signal-to-noise squared accounting for all annuli will be

$$\left(\frac{S}{N}\right)^2 = \sum_a \frac{N_a \gamma_t^2(\theta_a)}{\sigma_{\text{sh}}^2}. \tag{7.13}$$

So the signal-to-noise increases by a factor of the square root of the number of galaxies used to estimate the shear. In general, weak lensing therefore relies on measurements of many background galaxies in order to beat down the noise.

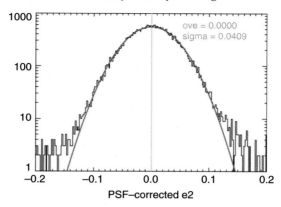

Figure 7.3 Distribution of ϵ_2 values in $0.1° \times 0.1°$ patches measured (Lin et al., 2012) in the Sloan Digital Sky Survey. The smooth curve is the best-fit Gaussian to the histogram, a fit which is seen to work well except for a few outliers at the extreme tails. Each patch has of order 130 galaxies in it, so the RMS of each patch, listed as sigma = 0.0409 in the figure, corresponds to an RMS in the ellipticities of individual galaxy of $(e_2)_{RMS} \simeq 0.04 \times \sqrt{130} = 0.45$. Most of this is due to shape noise, while some is due to measurement error. Recall that the definitions of shear and ellipticity lead to $\sigma_{sh} = (e_2)_{RMS}/2$ in the absence of measurement error.

7.3 Galaxy–Galaxy Lensing

Galaxies typically emit most of their light from a region within 5–10 kpc of their centers. Recall that the velocity of stars much further away from the center than this suggests that there is a lot of mass associated with galaxies that extends far beyond this visible region. Lensing offers another way of testing whether this is true.

Lensing offers an alternative path to estimating whether there is hidden mass, dark matter, associated with galaxies. The distortions in the shapes of background galaxies can be plotted as a function of their projected distance from the center of a foreground galaxy. The density profile of the galaxy, including all its mass, affects the shape of this shear vs. distance curve. So, *galaxy–galaxy lensing* – the distortions of shapes of background galaxies due to foreground galaxies – can be used to weigh in on the question of whether or not dark matter exists and, if so, how much resides in each galaxy. More generally, galaxy–galaxy lensing is a powerful way to understand the relation between the mass and luminosity of galaxies.

Let's begin by imagining a set of measurements of background galaxy shears such that the projected distance $R = D_L\theta$ from the foreground galaxy is much larger than 5–10 kpc, the visible extent of the foreground galaxy. In that case, if there were mass associated with only the visible part of the galaxy, then at these large distances we would be justified in approximating the expected signal to be that of a point mass. For a point mass, $\Sigma(\theta)$ is zero away from the origin, so the physical quantity defined in Eq. (7.9) would be

$$\left[\bar{\Sigma}(R) - \Sigma(R)\right]\bigg|_{\text{point mass}} = \frac{M}{\pi R^2}. \tag{7.14}$$

That is, we expect the shear signal in this case to fall off as R^{-2} as the background galaxies move further away in projected radius from the foreground galaxy. Compare this prediction with that from an assumed halo surrounding the galaxy, with an isothermal profile: $\rho \propto r^{-2}$. In that case, we showed in Eq. (3.25) that $\Sigma = \sigma^2/2GR$, and it is straightforward to show that $\bar{\Sigma}$ is twice as large as this, so the prediction is

$$\left[\bar{\Sigma}(R) - \Sigma(R)\right]\bigg|_{\text{isothermal halo}} = \frac{\sigma^2}{2GR}. \tag{7.15}$$

Fig. 7.4 illustrates the differences between these two predictions. If the galaxy's mass were indeed concentrated near the center as its light suggests, then the shear signal would be very different than if its mass were extended. The dashed curve shows the prediction for a point mass galaxy with the same amount of mass as the isothermal distribution shown within 10 kpc. The amplitude of the signal in that case would be significantly smaller than in the extended halo case. Even if we artificially boosted the amplitude as shown by the dotted points, the shape of the

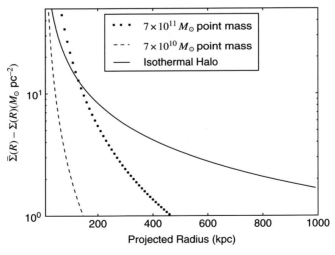

Figure 7.4 Predictions for the tangential shear of background galaxies produced by a foreground galaxy as a function of the projected distance between the two. The point mass, meant to approximate a galaxy with no dark matter, predicts a sharp fall-off in the signal, while the isothermal halo predicts a much slower decline. The two values of the point mass correspond to the mass of the isothermal halo within 10 kpc ($7 \times 10^{10} M_{\odot}$) and 100 kpc ($7 \times 10^{11} M_{\odot}$).

signal as a function of projected radius is significantly different from the isothermal profile.

We can estimate how difficult it would be to detect the difference between these two predictions using the signal-to-noise considerations of §7.2. In an annulus of radius θ_a and width $\Delta\theta$, the signal-to-noise in Eq. (7.13) for the isothermal halo reduces

$$\left(\frac{S}{N}\right)_a = \frac{[2\pi\bar{n}]^{1/2}}{\sigma_{sh}} \left(\frac{\Delta\theta}{\theta_a}\right)^{1/2} \left(\frac{\sigma^2}{2GD_L\Sigma_{cr}}\right) \qquad (7.16)$$

where we have substituted $R \rightarrow D_L\theta_a$ and $\bar{n}$ is the mean density of background galaxies per square radian. Note that the signal-to-noise per annulus does not vary much with the radius of the annulus (scaling as $\Delta\ln(\theta_a)$): the signal goes down as one gets further away, but the number of galaxies in the annulus goes up to compensate. You can show in Exercise (7.2) that the signal-to-noise in a given bin is of order 0.015 for realistic numbers. The way to increase this is to combine the signal from many foreground galaxies: a process called *stacking*. The signal-to-noise then goes up by the square root of the number of foreground galaxies observed. So observing $\sim 2 \times 10^4$ foreground galaxies would enable a detection at the 2-sigma level in each bin.

Fig. 7.5 shows such a stacked galaxy–galaxy lensing measurement using data from the Sloan Digital Sky Survey. The key feature to note is the gradual fall-off with radius: the signal drops by a factor of order ten moving from $R = 100$ kpc out to $R = 10^3$ kpc. Even putting aside the amplitude of the signal, this is in sharp contrast to the prediction of the "no dark matter" model, which predicts a fall-off that scales as R^{-2} and therefore would be a factor of a hundred smaller at $R = 1000$ kpc. A model wherein the only contribution to the mass comes from visible matter concentrated at the center is highly disfavored. Galaxy–galaxy lensing therefore provides one of the most compelling pieces of evidence for dark matter.

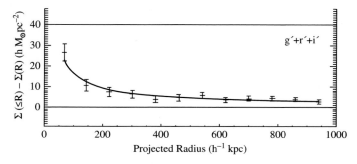

Figure 7.5 Measurements of stacked galaxy–galaxy lensing from the Sloan Digital Sky Survey (McKay et al., 2001). A total of 3.6 million background galaxies were used to compute the stacked signal induced by 31,000 foreground lens galaxies.

7.4 Implications for Dark Matter

Galaxy–galaxy lensing is important because it provides very direct evidence for dark matter and even a way of measuring the density profile of dark matter in galaxies. The robustness of the evidence lies in the fact that lensing probes the response of light to gravity, so provides complementary evidence to the velocities of stars depicted in Fig. 5.4, which speaks to the response of non-relativistic matter to gravity. It is not that difficult to construct a theory of modified gravity that changes the way matter responds to gravity and explains the flat rotation curves shown in Fig. 5.4. Most of those theories, however, do *not* affect the way light responds to gravity, so cannot account for the galaxy–galaxy lensing data depicted in Fig. 7.5. Therefore, galaxy–galaxy lensing data is one of the strongest arguments against modified gravity models that aim to solve the rotation curve anomaly.

It is easiest to illustrate this with an example, and the most famous example is MOdified Newtonian Dynamics (MOND). The idea is based on the insight (Milgrom, 1983) that Newton's Law $\vec{F} = m\vec{a}$ has been tested only at relatively large values of acceleration a, while stars rotating far from the centers of galaxies have very small centripetal accelerations. So, perhaps at low accelerations the law is different. In particular, MOND is the proposal that

$$\vec{F} = m\vec{a} \rightarrow \vec{F} = m\vec{a}\mu(a) \tag{7.17}$$

where $\mu(a)$ is a function of the acceleration that asymptotes to one for "normal" (i.e., large) values of the acceleration. So, in all experiments done on Earth, MOND makes the same predictions as Newtonian gravity (and general relativity). The precise form of μ is not important, only its limits

$$\mu = \begin{cases} 1 & a \gg a_0 \\ \frac{a}{a_0} & a \ll a_0 \end{cases} . \tag{7.18}$$

The exact value of the new parameter a_0 is to be determined by fitting to rotation curves, but we can get a sense of it by requiring that MONDian effects become important in the outskirts of galaxies where a is small. We expect a_0 to be of order the centripetal acceleration at the point where flat rotation curves are observed. There, velocities are of order 200 km s^{-1}, while distances are of order 10 kpc, so we expect $a_0 \sim 10^{-10}$ m s^{-2}. Further out than ~ 10 kpc, $a < a_0$, so in those regions Newton's law governing the rotation of stars changes

$$\frac{mMG}{r^2} = \frac{mv^2}{r} \rightarrow \frac{mMG}{r^2} = m\frac{v^2}{r}\left[\frac{v^2}{ra_0}\right] \tag{7.19}$$

where the term in square brackets is the MONDian correction at low a. Since both sides scale as r^{-2}, the velocity is predicted to be constant, not falling off in the Keplerian fashion. In just a few lines, we have described a theory that accounts for flat rotation curves, with predicted velocity $v = (MGa_0)^{1/4}$ independent of distance from the Galactic center. There is a simplicity to this idea that is amplified when attempting to fit rotation curves of many galaxies. With MOND, it is a simple matter of fitting the global parameter a_0 and introducing a single free parameter for each galaxy: the mass-to-light ratio. This is in sharp contrast to dark matter models, which fit rotation curves with three parameters: the mass-to-light ratio, and two parameters that describe the dark matter profile.

As presented, MOND is barely a theory. In order to make predictions for other phenomena, it needs to be promoted to the level of a theory that respects the invariance principles that are obeyed by general relativity. This has been done by a number of researchers with an interesting conclusion: the simplest way to generalize MOND is to embed it in a *scalar–tensor* theory of gravity and that, as we will see, light gets deflected in this class of theories in exactly the same way as in general relativity. Therefore, in these models, while MOND may account for flat rotation curves, it cannot account for galaxy–galaxy lensing. That is, this form of modified gravity affects massive particles (e.g., stars) differently than it does massless particles (e.g., the carriers of light, photons).

Scalar–tensor theories of gravity distinguish between two metrics that general relativity assumes are the same. The first is the one that enters into the Christoffel symbol in Eq. (2.28); this metric is the one that governs the way all particles move. It encodes the impact of curvature on the motion of particles. Then, there is a second metric: the one that is determined by the matter and energy. For example, for a point mass, this second metric is given by the expression in Eq. (2.33). General relativity implicitly assumes that these are the same; scalar–tensor theories posit that the two are different but proportional to one another with a space-time dependent proportionality coefficient often written as

$$g_{\mu\nu} = e^{2\phi} \tilde{g}_{\mu\nu}. \tag{7.20}$$

Here $g_{\mu\nu}$ is one of the metrics, $\tilde{g}_{\mu\nu}$ is the other, and $\phi(\vec{x}, t)$ is a scalar field that modulates the relation between them. Since general relativity contains the metric tensor, these models are called scalar–tensor models. The flexibility in them lies in the freedom to choose the dynamics that governs ϕ. Indeed, with a suitable choice for these dynamics, a scalar–tensor model can be constructed to reduce to MOND in the non-relativistic, low-acceleration limit. However, whatever ϕ does has no impact on the geodesics of massless photons. Small changes in space (dx^i; $i = 1$–3) and time ($dx^0 = dt$) for a massless particle are constrained by

$$ds^2 \equiv g_{\mu\nu}dx^\mu dx^\nu = 0. \tag{7.21}$$

Given the relation between the two metrics in scalar–tensor theories, though, as expressed in Eq. (7.20), the value of ϕ does not affect the value of ds^2 or therefore the relation in any coordinate system between dx and dt of a particular path. Therefore, the motion of massless particles predicted by any scalar–tensor theory will be identical to that predicted by general relativity.

Galaxy–galaxy lensing therefore offers up very strong evidence for dark matter.

7.5 Distances in Cosmology

Before turning to the issue of how and what can be learned about galaxy clusters from weak lensing, let us address an issue that has lurked behind many of the formulae derived so far: the definition of distance in an expanding universe. It is clear that the notion of distance in an expanding universe requires some care: is the distance between two objects the one when the light was emitted or the (larger) one at the later time when the light was detected, after the universe has expanded, so the objects are farther away from one another? Indeed, there are different measures of distance in cosmology that are used for different purposes.

The starting point to describe the expanding universe is to introduce the *scale factor* $a(t)$, which is increasing (hence the universe is expanding) and is normalized to be equal to one today. If one could place a ruler and measure the distance between any two objects today, that distance would be called the *comoving distance*, denoted χ. The physical distance, again with this hypothetical ruler, between the two objects in the past would have been equal to $\chi a(t)$. That is, the physical distance between these, or any, two objects was smaller than today by a factor of $a(t)$. All the changes due to expansion, therefore, are captured by the scale factor; the comoving distance between two objects remains constant throughout.[1]

Since the comoving distance remains constant during the expansion, it serves as a useful measure of the distance between any two objects. So, let's first calculate the comoving distance between a source that emits radiation at time t that is detected by us today at time t_0 (in cosmology, subscript $_0$ is often used to denote present-day values of time or quantities that change with time). In a small time dt, light travels a physical distance $c\,dt$ and therefore a comoving distance $d\chi = c\,dt/a(t)$. The comoving distance between us today at t_0 and a source that emitted light at an earlier time t therefore is

$$\chi(t) = c \int_t^{t_0} \frac{dt'}{a(t')}$$

$$= c \int_a^1 \frac{da'}{a'^2 H(a')}. \tag{7.22}$$

---

[1] There can be changes in the distance not caused by the expansion, but rather by local gravitational fields (such as a planet moving around a star or a galaxy moving in a galaxy cluster). These relative motions are called *peculiar velocities*.

The second equality follows after changing dummy variables from t' to a' and identifying the time derivative of the scale factor as the Hubble expansion rate

$$H \equiv \frac{da/dt}{a}. \tag{7.23}$$

Note that, generally, the expansion rate changes with time, so H is not a constant. Its value today, though, is called the *Hubble constant* and denoted H_0. The Hubble constant has been measured in a variety of ways, not all of which agree, so it is conventional to write

$$H_0 = 100\, h\, \text{km}\, \text{s}^{-1}\, \text{Mpc}^{-1} \tag{7.24}$$

with the uncertainty hidden in h (see shaded box for further details). Looking at Eq. (7.22), we see that the scale factor can be used as a proxy for time, so χ can be written as a function of time (the time at which the light was emitted) or as a function of a (the value of the scale factor at that time).

The Hubble Constant

The Hubble constant quantifies the rate at which the universe is expanding today. In principle, H_0 has units of time, but – as indicated in Eq. (7.24) – it is common to express it in units of velocity (km/s) divided by distance (Mpc). This makes sense physically, as H_0 quantifies how fast an object is moving away from us; at least at small distances, that recession speed is proportional to the distance, with H_0 serving as the proportionality constant. Note that the combination h^{-1} Mpc appears on the right-hand side of Eq. (7.24). When distances to objects are determined by their redshift, the value of h enters into the distance. In papers and books on cosmology, therefore, distances are often expressed in units of h^{-1} Mpc, and we will follow that convention.

The range of allowed values of h has tightened considerably since Hubble's initial estimate of $h \simeq 5$. Over the ensuing decades, a debate raged between two camps, one of which claimed $h = 0.5$ and the other $h = 1.0$. Current values hover in the range $h = 0.7$ with differences emanating from different techniques on the order of five percent.

There is another way to express the integral in Eq. (7.22) that defines the comoving distance; one that exploits the relationship between the scale factor and the redshift. The two are related in a trivial way. The redshift quantifies the amount by which the wavelength of light has expanded since it was emitted, but this is exactly the same thing quantified by the scale factor, so

$$a = \frac{1}{1+z}. \tag{7.25}$$

To be clear, this is a relation between the value of the scale factor when light we see today was emitted and the redshift of the observed light. Changing dummy variables in Eq. (7.22), we can write the comoving distance between us and an object at redshift z (see box on page 77) as an integral over the intervening redshifts:

$$\chi(z) = c \int_0^z \frac{dz'}{H(z')}. \tag{7.26}$$

Note that, at very low redshifts $z \ll 1$, the comoving distance reduces to $\chi(z) \to cz/H_0$. When Edwin Hubble first discovered the expansion, he measured the recession velocities of galaxies for which this relation was a very good approximation. The low redshift limit also conforms to our intuition: an object moving away from us with velocity cz has a spectrum redshifted by a factor of $1 + z$. So, in this limit, Hubble's discovery was that galaxies move away from us with a velocity proportional to their distance from us. As we see from Eq. (7.26), the relation between redshift and distance becomes more complicated at larger distances.

The distances used in the text so far – D_L, D_S, and D_{SL} – have *not* been comoving distances. Rather, we have been implicitly using what is called the *angular diameter distance*. It is defined as the distance that maintains the classic astronomical relation between the physical size of an object, l_{phys}, and the angle θ it subtends on the sky

$$D_A \equiv \frac{l_{\mathrm{phys}}}{\theta}. \tag{7.27}$$

To derive the angular diameter distance, think in comoving space: the angle then is set by ratio of the comoving size of the object and the comoving distance to the object, $\chi(z)$. But the comoving size is equal to the physical size divided by $a(t)$ or equivalently multiplied by $1 + z$. So the ratio that determines the angle is

$$\theta = \frac{l_{\mathrm{phys}}(1+z)}{\chi(z)}. \tag{7.28}$$

Comparing to the definition in Eq. (7.27), we see that

$$D_A(z) = \frac{\chi(z)}{1+z}. \tag{7.29}$$

Fig. 7.6 shows this distance as a function of redshift. Note that, at low redshifts, the angular diameter distance is indeed equal to cz/H_0, so it is only distant objects for which these more accurate formulae must be used.

To promote all of our previous results to be accurate at large redshifts, we can simply replace $D_L \to D_A(z_L)$ and $D_S \to D_A(z_S)$. The difference between the lens and the source requires a bit more care; you can show in Exercise (7.5) that

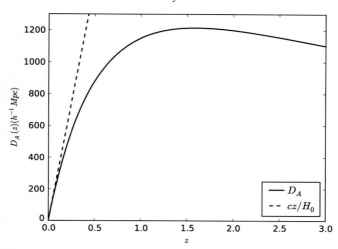

Figure 7.6 Angular diameter distance out to an object at redshift z. At low z, the linear approximation $D_A = cz/H_0$ works well, while at high redshift, the angular diameter distance decreases.

$$D_{SL} \rightarrow \frac{\chi(z_S) - \chi(z_L)}{1 + z_S}.$$ (7.30)

It is important to take a step back to think about the implications of these distance formulae. Each distance is an integral over the expansion rate in the past. This means that lensing observations, which are sensitive to these distances, have the potential to measure the expansion rate in the past. The expansion rate is tied to the density in the universe (see Exercise (7.4)), so determining how the expansion rate has changed with time speaks to what the universe is made of. More generally, gravity tends to lead to deceleration, and that holds in the case of the expanding universe, where the equation governing the evolution of the scale factor (again, see Exercise (7.4)) dictates that the expansion should be slowing down ($\ddot{a} < 0$). Observations from a number of different probes suggest otherwise: that the universe is accelerating, that $d^2a/dt^2 > 0$. This is a big puzzle, the solution to which may be a new substance dubbed *dark energy* or perhaps a change in the laws of gravity. We will return to this problem and some of the ways lensing addresses it in subsequent chapters.

7.6 Galaxy Clusters

Galaxy clusters are the largest bound objects in the universe. Fig. 7.7 shows an example: the galaxy cluster A2552. This is part of the Abell catalog of galaxy clusters (Abell et al., 1989) (hence the "A" prefix in its name). One of the criteria for selection in the Abell catalog is that the cluster must contain more than thirty

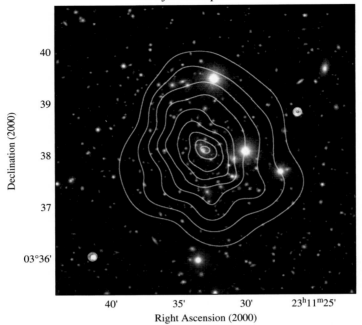

Declination (2000)

40

39

38

37

03°36'

40'   35'   30'   23^{h}11^{m}25'

Right Ascension (2000)

Figure 7.7 Emission from A2552 in X-ray and optical bands. The contours depict increasing X-ray flux towards the cluster center as measured by the Chandra satellite, while the colored image is based on optical observations with the University of Hawaii 2.2 meter telescope. The optical image reveals many individual galaxies in the cluster (which is at redshift $z = 0.3$), while the X-ray contours trace the hot gas that is concentrated in the cluster center. From Ebeling et al. (2010). (See color plates section.)

galaxies. This particular cluster has many more than this (94), as can be seen in the image. Generally, the more galaxies in the cluster, the more massive it is. The contours in Fig. 7.7 show the level of X-ray emission increasing towards the center. Clusters typically have even more mass in free electrons and protons (gas) than resides in the discrete galaxies. The gas is typically hot so the process $e^- + p \rightarrow e^- + p + \gamma$, *Bremsstrahlung*, leads to the emission of photons with energy of order the gas temperature; for A2552, the temperature is 8.7×10^7 K corresponding to an energy scale of $kT = 7.5$ keV. Generally, the temperature and X-ray luminosity correlate tightly with cluster mass, as you can estimate in Exercise (7.6).

Even this single illustration offers a glimpse into how interesting galaxy clusters are and raises a number of questions. What governs the formation and internal dynamics of clusters? What is their density profile? How is the gas distributed? What is the relation between the (dominant) dark matter, the constituent galaxies, and the gas? What is the mass of a cluster and how can it be determined? Gravitational lensing has emerged as one of the most promising tools to study these questions. We will focus on estimating the masses of clusters, so it is first worthwhile to take a look at why cluster masses are so important.

Consider Fig. 7.8, a cartoon version of a one-dimensional density field, $\delta(x)$, defined as the fractional difference between the density at a given location $\rho(x)$ and the mean density $\bar{\rho}$. Here, δ at any point is drawn from a Gaussian distribution with mean zero and variance σ. The shaded region in the figure shows that, as expected, most of the universe is filled with values of $\delta^2 < \sigma^2$. Clusters form in the rare excursions, those few places where $\delta \gg \sigma$.

This affords an interesting way to infer the value of σ, the quantity that quantifies the amplitude of the density fluctuations. We will see more of this in the next two chapters, but for now it is important to note that much of cosmology today is devoted to understanding the distribution of matter in the universe. Having a handle on σ, which quantifies the width of the distribution, is therefore extremely valuable. The number of rare excursions, which is related to the number of clusters, informs us about σ, the width of the distribution. For a Gaussian distribution, the number of regions with density above a threshold δ_c is proportional to $e^{-\delta_c^2/2\sigma^2}$. That is, the number of clusters is exponentially sensitive to the value of σ. Figure 7.9 displays this sensitivity as a function of cluster mass. The more massive the cluster, the bigger an outlier it is: that is, the value of δ_c/σ increases. Therefore, the

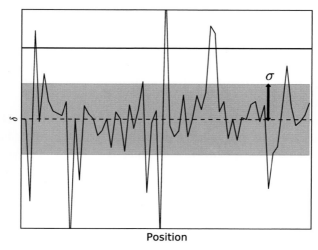

Position

Figure 7.8 Cartoon of the variation of the fractional density $\delta(x) \equiv (\rho(x) - \bar{\rho})/\bar{\rho}$ along one dimension, with the dashed line delineating positive from negative values. Shaded region delineates the square root of the variance of δ, its standard deviation. In the rare places where the density rises above a threshold, depicted by the horizontal solid line, large bound objects like galaxy clusters form.

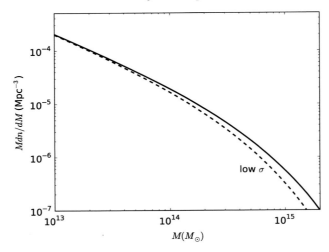

Figure 7.9 Predicted abundance of galaxy clusters as a function of their mass. The dashed curve shows the same prediction when the root mean square of the density fluctuations, σ, is lowered by 10 percent. Since the abundance is so sensitive to σ, the abundance at high mass drops precipitously, by more than a factor of two when $M > 10^{15} M_\odot$.

abundance of high-mass clusters is exponentially small. Measuring the number of clusters is a powerful way to infer σ.

One of the great successes of cosmology is the realization that the stark density contrasts found in the universe – great clusters of galaxies connected by long thin filaments separated by large voids – evolved from relatively smooth initial conditions. Billions of years ago, there were no galaxies, and the universe was smooth with $\sigma \ll 1$. Over the course of billions of years, the slightly overdense regions accreted more and more matter until they grew into the majestic structures observed today. This qualitative realization is quantified by the growth with time of the RMS density fluctuations, σ.

Imagine now taking the measurement of clusters to the next level by measuring the cluster abundance at many different redshifts. This would enable charting out $\sigma(z)$, which directly speaks to the rate at which σ grows with time and so touches on one of the most pertinent questions in all of cosmology. In Chapter 8, we will introduce the *growth function* instead of $\sigma(z)$, but they are targeted at the same phenomenon: the rate at which structure grows in the universe. We will see that the growth function depends on the underlying model and the parameters that describe it. So clusters empower us to measure the growth function, which in turn constrains cosmological parameters.

That is the good news. The more subtle news is that this program of inferring σ from the abundance of clusters requires us to know cluster masses extremely well. Looking back at Fig. 7.9, we see that the mass function at $M = 10^{14} M_\odot$ scales as $dn/dM \propto M^{-2}$ and is even steeper than that at higher masses. Now imagine a way of estimating masses that gets the correct answer on average (this is called an *unbiased estimator*) but has a scatter of 30 percent. If we want to estimate the number of clusters with a given mass, the scatter means that some of the many more clusters at lower mass will scatter into our number estimate. So the steepness of the mass function makes it particularly important to obtain accurate estimates of mass.

7.7 Cluster Lensing

Measurements of shapes of background galaxies enable a determination of the masses of clusters. Consider again the simple example of a cluster with a profile that falls off as $\rho \propto r^{-2}$. Until now, we have written isothermal profiles in terms of the velocity dispersion σ, so that we could rewrite Eq. (7.15) as

$$\gamma_t(\theta) = \frac{2\pi\sigma^2}{c^2} \frac{1}{\theta} \frac{D_{SL}}{D_S},$$
(7.31)

but we now encounter two problems with this notation. The first is trivial: we are already using σ to denote something completely different – the RMS fluctuations in the density field depicted in Fig. 7.8 – so do not want to confuse it with the σ here, which denotes velocity dispersion in the cluster. More importantly, as is clear from Fig. 7.7, how much mass a cluster contains depends on where its boundary is defined, and there is no clear delineation. The standard used in the literature is to define a radius r_Δ such that the mass density within that radius is equal to a large, dimensionless number Δ (typically set to 200 or 500) times the *critical density of the universe*:

$$\rho_c(z) \equiv \frac{3H^2(z)}{8\pi G}$$

$$= 2.77 \times 10^{11} h^2 \frac{M_\odot}{\text{Mpc}^3} \left(\frac{H(z)}{H_0}\right)^2.$$
(7.32)

The definition of both r_Δ and M_Δ then is

$$\frac{M_\Delta}{4\pi r_\Delta^3/3} = \Delta \rho_c(z).$$
(7.33)

This can be used to replace σ in the coefficient of the isothermal density profile; in particular you will show in Exercise (7.7) that

$$\sigma^2 = 2\pi G \left(\frac{M_\Delta^2 \Delta \rho_c(z)}{48\pi^2}\right)^{1/3}.$$
(7.34)

Therefore, the tangential shear induced by an isothermal cluster with mass M_Δ is

$$\gamma_t(\theta) = \pi \left(\frac{\Delta}{2}\right)^{1/3} \left(\frac{M_\Delta G H(z)}{c^3}\right)^{2/3} \frac{1}{\theta} \frac{D_{SL}}{D_S}. \tag{7.35}$$

Let's use Eq. (7.35) to get some feel for the numbers and to estimate how difficult it is to measure cluster masses with weak lensing. Putting in some fiducial numbers with $\Delta = 200$ leads to

$$\gamma_t(\theta) = 0.11 \frac{1'}{\theta} \left(\frac{M_{200}}{2 \times 10^{14} M_\odot}\right)^{2/3} \left(\frac{H(z)}{H_0}\right)^{2/3} \frac{D_{SL}}{D_S}. \tag{7.36}$$

The expected shear then is of order 10 percent at $\theta = 1'$ for a massive cluster and it scales as $M^{2/3}$. For a cluster with this mass at $z = 0.3$, $r_{200} \simeq 1$ Mpc, corresponding to $\theta \simeq 3'$. The question then becomes: suppose one can measure the shapes of all background galaxies within a radius of $3'$; how accurately can the mass be determined?

We see from simulated data shown in Fig. 7.10 that it should be possible to determine the mass of a single massive cluster with enough background galaxies. The signal is determined by Eq. (7.36), while the error bars represent the effect of shape noise: a random error with mean zero and variance equal to σ_{sh}^2/N_a, where N_a is the number of background galaxies in the annulus a. The example in Fig. 7.10 is a massive cluster with tangential shears measured in ten annuli with a background galaxy density of $\bar{n} = 10/$sq. arcminute. One other feature of the figure worth noting is that the first bin is at $\theta = 0.5'$. This is not unusual: although the signal increases towards the center, so does the noise because the radii of inner annuli, and therefore the number of background galaxies in them, is smaller than at larger distances. The gain in signal-to-noise therefore is minimal. More importantly, there are many cluster galaxies in the central region, and these can be mistaken for background galaxies. The cluster galaxies are not sheared by the cluster, so their mean tangential shear is zero. The ones that get mistaken as background galaxies therefore dilute the inferred signal. A simple fix is to exclude the central region, as hinted at in Fig. 7.10.

We could treat this problem as we did with galaxy–galaxy lensing and estimate the signal-to-noise, but it pays to introduce a slightly more rigorous formalism, the likelihood function. This is the probability of obtaining the data given the model. In this case, the data are the measured tangential shears in a set of annuli around the cluster center. Each measurement, $\gamma_{t,a}^{obs}$ in angular bin θ_a, is modeled to be the sum of the signal and noise. In this relatively simple model for the data, the difference between the observed tangential shears $\gamma_{t,a}^{obs}$ and the signal that depends on the mass, $\gamma_{t,a}(M_{200})$, is the noise, so should be drawn from a Gaussian distribution

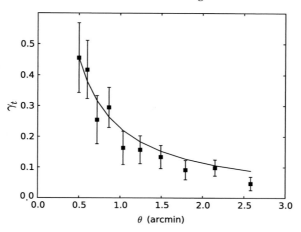

Figure 7.10 Simulated tangential shear in 10 radial bins induced by a foreground cluster at $z = 0.3$ with mass $M_{200} = 10^{15} M_\odot$. The background galaxies are all assumed to be at $z = 1$ and the shape noise per galaxy to be $\sigma_{sh} = 0.2$.

with mean zero and variance σ_{sh}^2 / N_a. That is, the probability of obtaining the data given the mass is

$$\mathcal{L}(M_{200}) = \prod_{a=1}^{N_{ann}} \frac{N_a}{(2\pi\sigma_{sh}^2)^{1/2}} \exp\left\{-\frac{1}{2} \left(\gamma_{t,a}^{obs} - \gamma_{t,a}(M_{200})\right)^2 \frac{N_a}{\sigma_{sh}^2}\right\}. \tag{7.37}$$

Given the data, we could compute this function, the likelihood, and transform it into constraints on the one free parameter, the mass M_{200} (see Exercise (7.8)). In general, the likelihood is a complicated function of its parameters, in this case M_{200}. If it were Gaussian, however, it would take the form $\mathcal{L} \propto e^{-(M-\bar{M})^2/2\sigma_M^2}$ where $\bar{M}$ is the best fit value, σ_M the one-sigma error, and the subscript $_{200}$ has been dropped for simplicity. In that case, the second derivative of the likelihood would determine the error. In particular,

$$\frac{1}{\sigma_M^2} = -\frac{\partial^2 \ln \mathcal{L}}{\partial M^2}. \tag{7.38}$$

You can derive a similar formula in Exercise (7.1) when the likelihood is simply $e^{-x^2/2}$. Consider the first derivative:

$$-\frac{\partial \ln \mathcal{L}}{\partial M} = \sum_{a=1}^{N_{ann}} \frac{\partial \gamma_{t,a}(M)}{\partial M} \left(\gamma_{t,a}^{obs} - \gamma_{t,a}(M)\right) \frac{N_a}{\sigma_{sh}^2}. \tag{7.39}$$

The second derivative acts on both the $\partial\gamma_{t,a}/\partial M$ in front and the term in parentheses. But we can neglect the first of these, for the remaining piece will be proportional to $\sum_{a=1}^{N_{ann}} \left(\gamma_{t,a}^{obs} - \gamma_{t,a}(M)\right)$, which will fluctuate around zero due to the noise. So the sum will be very small. That leaves only the second piece to contribute to the second derivative, and we are left with

$$\frac{1}{\sigma_M^2} = \sum_{a=1}^{N_{ann}} \left(\frac{\partial \gamma_{t,a}(M)}{\partial M} \right)^2 \frac{N_a}{\sigma_{sh}^2}. \tag{7.40}$$

This is quite a handy formula for it answers our initial question – how accurately do we expect the cluster mass to be extracted given a dataset – without actually requiring the data!

In the simple model we have been exploring, $\gamma_t \propto M^{2/3}$, so the derivative is $(2/3)\gamma_t/M$ and projected error on the mass reduces to

$$\frac{\sigma_M}{M} = \frac{3\sigma_{sh}}{2} \left[\sum_{a=1}^{N_{ann}} N_a \gamma_{t,a}^2(M) \right]^{-1/2}. \tag{7.41}$$

The fractional error on the mass is proportional to the shape noise, which makes sense: the greater the noise, the harder it is to extract the signal. If the angular bins are spaced logarithmically, so that the number of galaxies in each bin is proportional to $\theta \Delta\theta \propto \theta^2 \Delta \ln \theta$, then each bin contributes equally to the signal-to-noise, as γ_t (in this model) scales as θ^{-1}. Plugging in numbers, we find

$$\frac{\sigma_M}{M} = 0.24 \left(\frac{\sigma_{sh}}{0.2} \right) \frac{D_S}{D_{SL}} \left(\frac{H_0}{H(z_L)} \right)^{2/3} \left(\frac{2 \times 10^{14} M_\odot}{M_{200}} \right)^{2/3} \left(\frac{10}{(1')^2 \bar{n}} \right)^{1/2}$$
$$\times \left(\frac{10}{N_{ann}} \right)^{1/2} \left(\frac{0.2}{\Delta\theta/\theta} \right)^{1/2}. \tag{7.42}$$

Eq. (7.42) gives us an estimate for the accuracy with which cluster masses can be measured with weak lensing. The shape noise is difficult to reduce much lower than 0.2. The next two factors incorporate the redshift dependence of the signal, also shown in Fig. 7.11. As the background galaxies get closer to the cluster, the signal gets weaker (scaling as D_{SL}) so the error blows up. We will return to this factor shortly; for now, let's estimate that it is roughly equal to two. The terms on the second line are coupled (you can get more annuli by taking $\Delta\theta$ smaller, so those two terms are roughly $(2/\int d\theta/\theta)^{1/2} = (2/\ln(\theta_{max}/\theta_{min}))^{1/2}$ which is roughly equal to unity. Obtaining tighter constraints on the mass then boils down to getting more than ten galaxies per square arcminute or considering larger clusters.

The results from one of the more ambitious programs to measure cluster masses are shown in Fig. 7.12. The clusters in this sample were quite massive, with $M_{200} > 6 \times 10^{14} M_\odot$, and the background density was about 20 per square arcminute. So, by our estimate above, the fractional error on the masses should have been about $0.24 \times 2 \times (2/6)^{2/3} \times (10/20)^{1/2} = 0.15$. This is a bit lower than the actual errors shown in Fig. 7.12. The errors we've estimated so far are

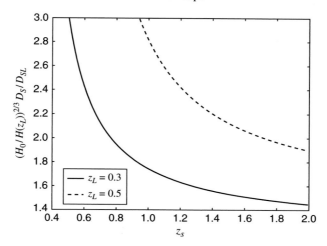

Figure 7.11 The redshift-dependent factor in the projected error σ_M/M for a cluster at z_L as a function of the redshift of the background galaxies. Background galaxies close to the lens ($z_S \to z_L$) produce a very small signal and so large errors on the cluster mass.

statistical errors, solely driven by the shape noise and the number of background galaxies. There are systematic errors that are just as important in this case and tend to increase the overall uncertainty. We have already mentioned one such error: the leakage of cluster galaxies into the background sample. Another can be gleaned from Fig. 7.11. The signal is quite sensitive to the redshifts of the background galaxies. The most powerful way to determine a galaxy's redshift is to take its spectrum, but this takes quite a bit of time and is therefore impractical given that we require high densities to beat down the statistical noise. The alternative, then, is to measure the flux from a galaxy in several different wavelength bands and try to infer the redshift from these *photometric* data. The ensuing photometric redshifts can be quite accurate (to better than 10 percent) but they do lead to an important source of error in the measurement of cluster masses. Another systematic error that plagues cluster mass measurements is the determination of the true center of mass of the cluster, around which the tangential shear will be measured. The realistic errors shown in Fig. 7.12 account for these (and others) and therefore are slightly worse than expected from Eq. (7.42).

7.8 Mass Maps

Section 7.7 contains a discussion of estimating the mass of the cluster, or more precisely projects how tightly a parameter, M_{200} say, that characterizes the cluster mass can be determined. As such, it is a *parametric* analysis. One can imagine generalizing this in several ways: more than one parameter is often needed to describe the

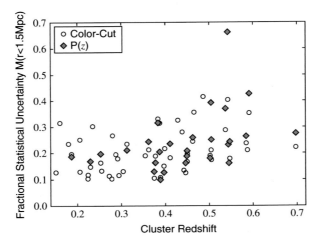

Figure 7.12 Fractional error on cluster masses as a function of their redshift from the "Weighing the Giants" program (Applegate et al., 2014). The different symbols correspond to two different ways of inferring the redshift distributions of the background galaxies.

density profile of a galaxy cluster. We focused on a particularly simple profile that had just one free parameter. Even if we had introduced several other parameters, we were implicitly assuming that the mass distribution was spherically symmetric, so that is another direction one might move in to generalize the result. Perhaps the most extreme generalization is to assume nothing about any symmetry or any parametric form for the density but to simply try to extract the mass directly from the measured shears. It is this general idea of a non-parametric mass map that we would like to explore here.

The most direct estimate of mass is the convergence κ. When the distortions are due to a single lens, then κ is defined in Eq. (2.65) to be the surface density divided by the critical surface density. More generally, we will see in the next chapter that $\kappa(\vec{\theta})$ is a weighted integral along the line of sight of the density field $\delta(\vec{x})$. We can even imagine breaking up the background galaxies into multiple redshift bins and obtaining κ maps for each background bin, thereby learning about the 3D mass distribution. But here we will limit ourselves to the 2D κ map.

To begin, consider again Eq. (4.5), which relates both the convergence κ and the two components of shear $\gamma_{1,2}$ to the projected gravitational potential. These relationships are of the form $\partial^2 \Phi / \partial \theta_i \partial \theta_j$. So if it were simple to take the inverse of these second derivatives, then Φ could be recaptured from the shear, and then κ extracted by differentiating Φ twice again. This process is most convenient if we deal with the Fourier transforms of the fields. The only thing hard about Fourier transforms is remembering the conventions, so our conventions are described in the shaded box.

Fourier Transform Conventions

Any field can be expressed in terms of its Fourier transform. For the rest of this chapter and in the remaining two chapters, we will use *continuous* Fourier transforms. Instead of decomposing a function $f(x)$ as

$$f(x) = \sum_k f_k \cos(kx) \tag{7.43}$$

we will use the continuous analog

$$f(x) = \int \frac{dk}{2\pi} \tilde{f}(k) e^{ikx}. \tag{7.44}$$

Most of our examples will be with functions of a two-dimensional vector. So, for example, we will decompose the convergence as

$$\kappa(\vec{\theta}) = \int \frac{d^2 l}{(2\pi)^2} \tilde{\kappa}(\vec{l}) e^{i\vec{l}\cdot\vec{\theta}}. \tag{7.45}$$

This simple relation encodes the basic features of all Fourier transforms, but in a way that will prove useful for many aspects of lensing, so it pays to enumerate them:

- A given function can be expressed in real space (here $\kappa(\vec{\theta})$) or in terms of its Fourier coefficients, here $\tilde{\kappa}(\vec{l})$. When using the latter, we often say we are working in "Fourier space."
- The basis functions are not $\sin(\vec{l} \cdot \vec{\theta})$ or $\cos(\vec{l} \cdot \vec{\theta})$ but rather the closely related $e^{i\vec{l}\cdot\vec{\theta}}$, which depend on the (in this case) two-dimensional vector, $\vec{l}$, sometimes called the conjugate to the real space variable $\vec{\theta}$.
- The low-l modes correspond to basis functions varying slowly as a function of $\vec{\theta}$; hence they are called *large-scale* modes. Similarly, the high-l modes carry information about the function on small scales.
- The sum over the basis functions in continuous space becomes a two-dimensional integral over $d^2 l$, and the factor of $(2\pi)^2$ is a (convenient) normalization choice.
- To go back and forth between real space and Fourier space, the identity

$$\int d^2 l\, e^{i\vec{l}\cdot\vec{\theta}} = (2\pi)^2 \delta_D(\vec{\theta}) \tag{7.46}$$

is very useful. Using this you can show (Exercise (7.10)) that

$$\tilde{\kappa}(\vec{l}) = \int d^2\theta\, e^{-i\vec{l}\cdot\vec{\theta}} \kappa(\vec{\theta}). \tag{7.47}$$

With this tool in hand, we can easily manipulate the relation between κ and γ to obtain a mass map (κ) from the shears. Consider the Fourier transform of the convergence definition in Eq. (4.5):

$$\int d^2\theta e^{-i\vec{l}\cdot\vec{\theta}}\kappa(\vec{\theta}) = \frac{1}{2c^2}\int d^2\theta e^{-i\vec{l}\cdot\vec{\theta}}\left(\frac{\partial^2\Phi}{\partial\theta_x^2} + \frac{\partial^2\Phi}{\partial\theta_y^2}\right). \tag{7.48}$$

The left-hand side is equal to $\tilde{\kappa}(\vec{l})$. We can treat the terms on the right-hand side by integrating by parts. For example, the first term contains

$$\begin{aligned}
\int d^2\theta e^{-i\vec{l}\cdot\vec{\theta}}\frac{\partial^2\Phi}{\partial\theta_x^2} &= -\int d^2\theta \frac{\partial(e^{-i\vec{l}\cdot\vec{\theta}})}{\partial\theta_x}\frac{\partial\Phi}{\partial\theta_x} \\
&= \int d^2\theta \frac{\partial^2(e^{-i\vec{l}\cdot\vec{\theta}})}{\partial\theta_x^2}\Phi(\vec{\theta}) \\
&= -l_x^2\tilde{\Phi}(\vec{l})
\end{aligned} \tag{7.49}$$

where the first and second lines follow by integrating by parts once and then a second time. In both cases, the boundary terms vanish because we can set $\Phi \to 0$ far from the region of interest. The final equality again follows after differentiating the exponential twice and then recognizing the remaining integral as the definition of the Fourier transform of the projected gravitational potential. Applying the same set of steps to the second term leads to an algebraic equation that relates the potential and the convergence in Fourier space:

$$\tilde{\kappa}(\vec{l}) = \frac{-l^2}{2c^2}\tilde{\Phi}(\vec{l}). \tag{7.50}$$

There are two lessons to be gleaned from this expression. First, derivatives in real space appear as powers of the conjugate variable in Fourier space. Second, and related, but incredibly useful for checking calculations is the way that dimensional analysis works with angular distances. Although we ordinarily think of angular distances θ as being dimensionless (expressed in radians), it is quite convenient to recognize that the conjugate variable l has inverse dimensions (radians^{-1} if you like). So while $\kappa(\vec{\theta})$ is dimensionless, $\tilde{\kappa}(\vec{l})$ has dimensions of θ^2 or radians squared. This will often help us keep track of the various factors of l. Meanwhile, the angular Laplacian of the projected potential is related to κ, so Φ/c^2 has angular dimensions of radians^2. Therefore, $\tilde{\Phi}/c^2$ has dimensions of radians^4, and we do indeed need the factor of l^2 in front of Eq. (7.50) simply to get the dimensions correct.

Similarly, we can transform the shear-potential relationship in Eq. (4.5) to Fourier space, leading to

$$\begin{aligned}
\tilde{\gamma}_1(\vec{l}) &= \frac{-l_x^2 + l_y^2}{2c^2}\tilde{\Phi}(\vec{l}) \\
\tilde{\gamma}_2(\vec{l}) &= \frac{-l_x l_y}{c^2}\tilde{\Phi}(\vec{l}).
\end{aligned} \tag{7.51}$$

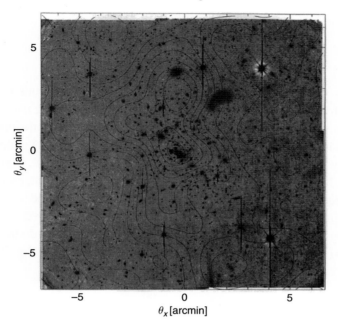

Figure 7.13 Mass map of the cluster Abell 2218 (Squires et al., 1996). The contours delineate equal values of κ, with the largest located near the center at the position $\theta_x = \theta_y = 0$. The black images are the galaxies in the cluster, also concentrated near the center.

Substituting for $\tilde{\Phi}$ from Eq. (7.50) into each of these equations, multiplying the first by $l_x^2 - l_y^2$ and the second by $2l_x l_y$, and then adding them leads to

$$\tilde{\kappa}(\vec{l}) = \frac{(l_x^2 - l_y^2)\tilde{\gamma}_1(\vec{l}) + 2l_x l_y \tilde{\gamma}_2(\vec{l})}{l^2}. \tag{7.52}$$

This expression holds everywhere except when $l = 0$, which corresponds to a Fourier mode with no variation, i.e., a constant. So this reconstruction is correct up to an overall constant value of κ. This is the so-called *mass sheet degeneracy*. Apart from this, knowing $\tilde{\kappa}(\vec{l})$ is equivalent to knowing $\kappa(\vec{\theta})$; that is, to constructing a mass map. So Eq. (7.52) can be used to transform the measured shears in a survey into a mass map of the region. An example is shown in Fig. 7.13.

There is one subtlety used to make the map shown in Fig. 7.13: the shears contain not only the signal but also noise in form of shape noise. So far, we have characterized the noise in terms of its variance, Eq. (7.12). A slightly more general expression would incorporate the fact that the shape noise in one galaxy is (generally) uncorrelated with that of another, so taking γ_1 as an example,

$$\langle \gamma_{1,i}\gamma_{1,j} \rangle = \langle \gamma_{1,i} \rangle \langle \gamma_{1,j} \rangle = 0 \qquad (i \neq j). \tag{7.53}$$

This can be combined with Eq. (7.12) into a single equation

$$\langle \gamma_{1,i} \gamma_{1,j} \rangle = \delta_{ij} \sigma_{\text{sh}}^2 \tag{7.54}$$

where δ_{ij} is the Kronecker delta, which restricts the fluctuations to be non-zero only when focusing on one galaxy. One can imagine summing over all N galaxy shears in a given pixel to obtain a single estimate of the shear in the pixel with error $\sigma_{\text{sh}}/\sqrt{N}$. This could also be captured by Eq. (7.54) where the i, j denote pixels, not galaxies. All this is a way of expressing the noise in real space. You can show in Exercise (7.11) that the variance in Fourier space is also constant, independent of l. The signal squared, on the other hand, scales as l^{-2} in our simple isothermal sphere model, so on small scales (large l), the noise will dominate over the signal. Any map-making algorithm must account for this. A simple way to do this is to cut out all the high-l modes; that is what was done to make the map in Fig. 7.13.

Suggested Reading

Since so much of weak lensing is focused on extracting signal from noise, I cannot help but recommend Silver (2012), which of course does not cover gravitational lensing but does have chapters about a wide variety of topics and even hints at the likelihood introduced here.

The story of MOND and its successors is extremely interesting. It begins with the paper by Milgrom (1983). A very clear summary of attempts to promote MOND to a full relativistic theory is detailed in the first few pages of Bekenstein (2004), which also introduces a model that accounts for lensing and reduces to MOND. A clear defense by two of its proponents is given in Sanders and McGaugh (2002). I do not find these arguments compelling; for more on why, see Dodelson (2011).

The way that large objects such as clusters form is a fascinating topic that has spawned a lot of interesting theoretical work. A superb guide to this work, and an excellent overall guide to structure formation in cosmology, is given by Zentner (2007). The cluster mass determinations summarized in Fig. 7.12 come from one (Applegate et al., 2014) of a series of careful papers by the *Weighing the Giants* collaboration. The systematic errors alluded to in §7.6, and others, are explained very well there.

The best guide to cosmological distances is the short, but comprehensive article posted by David Hogg (Hogg, 1999).

Techniques for producing mass maps from galaxy shape measurements go back to Kaiser and Squires (1993), whose Kaiser–Squires method remains in wide use and is similar to what is described in §7.8. The map in Fig. 7.13 was made using a slightly modified version of the Kaiser–Squires method.

Exercises

7.1 It is clear from Exercise (4.8) that the curvature of the χ^2 determines the strength of a constraint. So, one way to estimate the expected signal-to-noise of a measurement is to compute the curvature of this χ^2: the steeper the χ^2 as a function of the parameters, the more constraining will be the data, and therefore the larger is the signal-to-noise. Consider the case discussed in §7.2 where the signal $\gamma_t(\theta_a)$ is known. Introduce a parameter A that multiplies γ_t so is equal to one. An estimate of the signal-to-noise is equivalent to an estimate of how well A will be constrained. Compute the curvature of

$$\chi^2(A) = \sum_a \frac{N_a(\gamma_t^{\text{meas}}(\theta_a) - A\gamma_t(\theta_a))^2}{\sigma_{\text{sh}}^2} \tag{7.55}$$

as $F = (1/2)d^2\chi^2/dA^2$. Show that this estimate of the signal-to-noise squared agrees with Eq. (7.13).

7.2 Estimate the signal to noise from galaxy–galaxy lensing (using Eq. (7.16)) in an annulus of radius $\theta = 500$ kpc with width 100 kpc. Assume that the foreground galaxy is a distance $D_L = 1$ Gpc away and the source 1.7 Gpc. Take the velocity dispersion to be $\sigma = 120$ km sec^{-1} and the background galaxy density to be $\bar{n} = 3$ per square arcminute. This value of $\bar{n}$ is appropriate for the galaxy–galaxy lensing detection in the Sloan Digital Sky Survey depicted in Fig. 7.5, but surveys such as the Dark Energy Survey and especially the Large Scale Synoptic Survey (LSST) will have background galaxies a factor of 3–10 times larger than this.

7.3 Compute the angular diameter distance to an object at $z = 1$. Assume that

$$H(z) = H_0 \left[\Omega_m(1 + z)^3 + (1 - \Omega_m) \right]^{1/2} \tag{7.56}$$

and set the parameter that quantifies the matter in the universe, $\Omega_m = 0.31$.

7.4 The expression for the Hubble expansion rate in Exercise (7.3) arises from the Friedmann equation, which governs the rate at which the universe expands:

$$H^2 = \frac{8\pi G}{3c^2} \rho \tag{7.57}$$

where ρ is the smooth energy density of the universe. Show that the Friedmann equation reduces to Eq. (7.56) when the density is composed of matter and a cosmological constant: $\rho = \rho_m + \rho_\Lambda$. Use the facts that:

- The density of matter, ρ_m, falls as the universe expands, scaling as the volume a^{-3}, while the density associated with the cosmological constant is independent of a.

- The matter density and cosmological constant density today sum to the critical density $\rho_{\rm cr} \equiv 3c^2 H_0^2 / 8\pi G$. Note that this definition is consistent with Eq. (7.32) with the understanding that $\rho_{\rm cr}$ without any dependence on z refers to its current value.
- The ratio of the matter density to the critical density today is defined as $\Omega_m \equiv \rho_{m,0}/\rho_{\rm cr}$.

7.5   In comoving coordinates, draw the triangle that has at its base a source of a given size at redshift z_S and its apex at z_L. Use this triangle to verify Eq. (7.30).

7.6   Apply the virial theorem to estimate the mass of a cluster. A virialized system has kinetic energy equal to $-1/2$ times its potential energy. The total kinetic energy per free proton is $\langle v^2 \rangle/2$. Express the thermally averaged $\langle v^2 \rangle$ in terms of the temperature and the proton mass. The total potential energy per proton within a radius R can be estimated by assuming that the cluster is spherical with a density profile that falls off as r^{-2}. Show that this leads to the estimate

$$M = \frac{5k_B T R}{m_{\rm pr} G}. \tag{7.58}$$

Estimate this for the cluster depicted in Fig. 7.7.

7.7   Show that the coefficient in Eq. (7.34) follows from the definitions of r_Δ and M_Δ. Express the mass in terms of the velocity dispersion squared.

7.8   Consider a dataset D and a set of parameters used to model the data, λ. The probability of obtaining both the data and the parameters is equal to:

$$P(D \cap \lambda) = P(D|\lambda)P(\lambda) \tag{7.59}$$

where $P(D|\lambda)$ is the conditional probability of obtaining the data given the model parameters λ; this is what we called the likelihood and $P(\lambda)$ is called the *prior*, the expected range of λ given previous data or even theoretical prejudices. Using the same logic as above, write the joint probability in terms of $P(D)$ and $P(\lambda|D)$, and then equate your expression with the one in Eq. (7.59) to obtain an expression for $P(\lambda|D)$. This is what we want: the probability of the parameters given the data. It is called Bayes' Theorem. This probability then can be used to infer, e.g., 68 percent confidence levels. These "one-sigma" upper and lower limits λ_+ and λ_- are obtained by finding the values on either side of the maximum of $P(\lambda|D)$ such that $P(\lambda_+|D) = P(\lambda_-|D)$ and

$$\int_{\lambda_-}^{\lambda_+} d\lambda \, P(\lambda|D) = 0.68 \tag{7.60}$$

assuming that the probability – like all probabilities – is normalized so that the integral over all λ is equal to one. What is the value of $\lambda_\pm$ if $P(\lambda|D)$ is a Gaussian with width equal to σ_λ?

7.9 Plot the likelihood for M_{200} in the example shown in Fig. 7.10. The simulated dataset consists of ten points.

θ (arcmin)	γ_t^{obs}
0.5	0.455
0.6	0.209
0.72	0.342
0.86	0.233
1.04	0.186
1.24	0.188
1.49	0.206
1.79	0.127
2.15	0.089
2.58	0.107

Recall that the number density of background galaxies is assumed to be ten per square arcminute; this factors into the noise for each bin, which also depends on $\sigma_{sh} = 0.2$. You should be able to compute that the ratio D_{SL}/D_L that appears in the model is equal to 0.63 for the cluster at $z = 0.3$ and the background galaxies at $z = 1$; similarly take $H(z = 0.3)/H_0 = 1.17$. Determine the best fit value of M_{200} and the 1-sigma error bars, as outlined in the previous problem. Compare these error bars with those estimated in Eq. (7.42).

7.10 Show that Eq. (7.47) holds by multiplying Eq. (7.45) by $e^{-i\vec{l}'\cdot\vec{\theta}}$ and integrating over $d^2\theta$.

7.11 The continuous space version of Eq. (7.54) is

$$\langle \gamma_1(\vec{\theta})\gamma_1(\vec{\theta}')\rangle = (2\pi)^2 \delta_D(\vec{\theta} - \vec{\theta}')\frac{\sigma_{sh}^2}{N}\Delta^2 \qquad (7.61)$$

where Δ is the angular extent of a pixel with N galaxies in it. Show that the corresponding expression in Fourier space is

$$\langle \tilde{\gamma}_1(\vec{l})\tilde{\gamma}_1(\vec{l}')\rangle = \delta_D(\vec{l} + \vec{l}')C_l^{noise} \qquad (7.62)$$

where the *angular power spectrum* of the noise C_l^{noise} is independent of l and equal to $\frac{\sigma_{sh}^2}{N}\Delta^2$.

8

Cosmic Shear

Cosmic shear is responsible for the distortions in photon paths induced by the cosmos: the inhomogeneous universe through which photons propagate after they are emitted by sources such as galaxies. While the shear in a given direction can be induced primarily by a single object such as a massive galaxy cluster, more often it is the variations in the density, and therefore the gravitational potential, along the line of sight that combine to produce the shear. Cosmic shear then offers a way of learning about the distribution of matter in the universe. Cosmological models make predictions for how matter is distributed, so cosmic shear is a potentially powerful way of discovering the true model of the universe, or equivalently of answering age-old questions, such as "How old is the universe?", "How did we get here?" (or the more modern version: "How did structure form?"), and "What is the universe made of?"

Armed with this motivation, let's collect the relevant definitions from earlier chapters. Eq. (2.55) expresses the 2D projected gravitational potential as an integral along the line of sight of the 3D gravitational potential ϕ from a source at distance D_S

$$\Phi(\vec{\theta}) \equiv \frac{2}{D_S} \int_0^{D_S} dD_L \, \phi(x^i = D_L \theta^i, D_L; t = t_0 - D_L/c) \frac{D_{SL}}{D_L}. \tag{8.1}$$

Recall that here x^i denotes the two components of the 3D position transverse to the line of sight and that the potential is evaluated at an epoch t determined by the time at which the photon passed through that point. It is this potential that determines the components of cosmic shear. In particular, repeating Eq. (4.4)

$$\kappa \equiv \frac{1}{2c^2} \left(\frac{\partial^2 \Phi}{\partial \theta_x^2} + \frac{\partial^2 \Phi}{\partial \theta_y^2} \right)$$

$$\gamma_1 \equiv \frac{1}{2c^2} \left(\frac{\partial^2 \Phi}{\partial \theta_x^2} - \frac{\partial^2 \Phi}{\partial \theta_y^2} \right)$$

$$\gamma_2 \equiv \frac{1}{c^2} \frac{\partial^2 \Phi}{\partial \theta_x \partial \theta_y}. \tag{8.2}$$

Using these relations, we can express the shears, which, we know from Chapter 6 can be inferred from measurements of galaxy shapes, in terms of the gravitational potential. As we will see, this is very powerful because theories make predictions for the nature of the fluctuations in the gravitational potential.

We saw in §7.8 that the relation between the shear observables and the potential is very simple in Fourier space:

$$\tilde{\gamma}_1(\vec{l}) = \frac{-l_x^2 + l_y^2}{2c^2} \tilde{\Phi}(\vec{l})$$

$$\tilde{\gamma}_2(\vec{l}) = \frac{-l_x l_y}{c^2} \tilde{\Phi}(\vec{l}). \tag{8.3}$$

Similarly, there is an algebraic relation between the potential and the convergence in Fourier space

$$\tilde{\kappa}(\vec{l}) = \frac{-l^2}{2c^2} \tilde{\Phi}(\vec{l}). \tag{8.4}$$

There is one more relation that will prove useful: that between the gravitational potential and the over-density field

$$\delta(\vec{x}, t) \equiv \frac{\rho_m(\vec{x}, t) - \bar{\rho}_m(t)}{\bar{\rho}(t)}, \tag{8.5}$$

which quantifies the fractional change in the energy density at any point from the mean density $\bar{\rho}_m$ at that time. This fractional over-density is related to the 3D gravitational field via Poisson's equation, suitably generalized to cosmology:

$$\nabla^2 \phi(\vec{x}, t) = \frac{4\pi G}{c^2} \bar{\rho}_m(t) a^2(t) \delta(\vec{x}, t), \tag{8.6}$$

where the factor of c^2 accounts for the fact that the density $\bar{\rho}_m$ is the energy density, not the mass density. The Laplacian here is with respect to comoving coordinates whose difference defines the comoving distances described in §7.5. The factors of a are important and enable us to see clearly the resolution to an interesting cosmological dilemma. On the one hand, the universe is expanding, so we might expect matter to get more dilute and the gravitational wells ϕ to decay with time. On the other hand, an over-dense region attracts more matter into its vicinity because of gravity, so we might expect potential wells to get deeper with time as more and more matter accretes into an over-dense region. Which effect wins? It turns out to depend on the dominant form of energy in the universe: if it is non-relativistic matter, as it has been for most of the universe's history, then the two effects delicately balance, and potential wells remain constant with time. Then, the right-hand side of Eq. (8.6) must be independent of time. The mean matter density $\bar{\rho}_m$ decreases as the universe expands: it scales as the volume or as a^{-3}. Therefore, in this situation where non-relativistic matter is the dominant form of energy, the overdensity δ grows as a. This is a quantitative way of expressing the growth function described qualitatively on page 148: it simply grows as a. Other possibilities are considered in Exercise (8.3).

To sum up, shear is produced by the projected gravitational potential Φ, which is a line-of-sight integral of the 3D potential, which in turn can be related to the matter density field. We will be interested in the statistics of these fields, as they inform us about the underlying cosmology, so we begin with a section on the basic cosmological model and then move to a section on the statistics of δ in that model, with the understanding that its statistics can be extracted from measurements of shear.

8.1 Standard Cosmological Model

The universe is expanding, a phenomenon that is encoded in the scale factor $a(t)$, which multiplies all distances, is increasing with time, and normalized to be equal to one today. As noted in Eq. (7.25), the scale factor is inversely related to the redshift so a given epoch can be identified by either the time t, the scale factor at that time $a(t)$, or the redshift that quantifies the wavelength-stretching of any photons emitted from an object at that time (see again the box on page 77).

Until now we have used one part of Einstein's theory of general relativity: the laws that govern the motion of particles moving in a space with a particular metric. The second part of general relativity is the idea that the metric itself is determined by the matter and energy in the region of interest. This part is codified by Einstein's equations that relate geometry to energy. One application of Einstein's equations is to the smooth universe at large. In that (simple) case, they reduce to a simple differential equation for $a(t)$, an equation known as the Friedmann equation:

$$\frac{\dot{a}}{a} = \left[\frac{8\pi G}{3c^2} \rho \right]^{1/2} \tag{8.7}$$

where ρ is the energy density. Qualitatively, this says that the universe is expanding ($\dot{a} > 0$) at a rate proportional to the square root of the energy density.

Before reporting on the census of the constituents of the universe that comprise the total energy density, it is worth (re)-introducing some notation. The rate on the left-hand side of the Friedmann equation is called the *Hubble rate* $H(t) \equiv \dot{a}/a$, and its value today is denoted H_0 (see box on page 143). The value of the density today, then, is the one that satisfies Eq. (8.7) with the left-hand side set equal to H_0. This value is called the *critical density* of the universe (see also Exercise (7.4)):

$$\rho_{cr} \equiv \frac{3H_0^2 c^2}{8\pi G}. \tag{8.8}$$

We will express densities of the constituents of the universe in units of ρ_{cr}. For example, the mean density of all matter in the universe today is

$$\Omega_m \equiv \frac{\bar{\rho}_{m,0}}{\rho_{cr}}. \tag{8.9}$$

Today, the major constituents of the universe are ordinary matter in the form of hydrogen, helium, and other elements; dark matter, a small fraction of which is comprised of light particles called *neutrinos*; and a mysterious substance known as *dark energy*. In addition, a small fraction of the density is comprised of photons in the form of the Cosmic Microwave Background (CMB). The density of matter was larger in the past because distances and therefore volumes were smaller. As mentioned above, we expect then that the matter density scales as a^{-3}, that is as the volume. The energy density of the cosmic microwave background was also larger in the past, but by an additional factor of a^{-1}. This captures the energy per photon, which, since it is inversely proportional to wavelength, was larger in the past by a factor of $1/a$. Today, the energy density of the CMB though is much smaller than that of other constituents so can be neglected. Fig. 8.1 shows the evolution of several components of the density as a function of redshift.

To define this model of the universe then, we need to specify the following parameters:

- *Baryon Density, Ω_b:* This includes all protons, neutrons, and electrons in whatever form: hydrogen, helium or any other elements. I.e., the name *baryons* is used in this context to denote all ordinary matter. Measurements, most precisely from the CMB, fix $\Omega_b = 0.049$ at the percent level.
- *Matter Density, Ω_m:* This includes not only baryons but also any dark matter. Evidence from a variety of observations, including the set from cosmic shear that we are gearing up to discuss, fix the total matter density to be $\Omega_m = 0.315$, again with percent level uncertainty.
- *Hubble constant, H_0:* Perhaps it would be useful again to re-scan the box on page 143, but the bottom line is that current measurements suggest $H_0 = 70$ km s^{-1} Mpc^{-1} with measurements differing at the 5% level.
- *Dark Energy Density:* Here things get more complicated because we do not know what dark energy is. As you can work through in Exercises (8.1) and (8.2), you will see that it must produce an accelerating universe and therefore

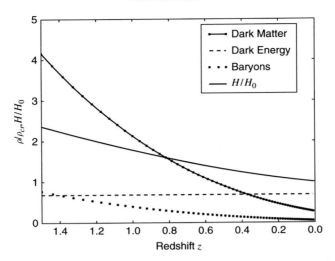

Figure 8.1 Evolution of the different components of the density as a function of redshift. Time flows from left to right, with today corresponding to $z = 0$. Apart from the dark energy, the energy density of the constituents of the universe was larger in the past due to the smaller volumes. The Hubble rate scales as $\rho^{1/2}$ so was also larger in the past.

must be roughly constant with time. The simplest model is that the dark energy is a *cosmological constant*, which has $\rho_{de} = $ constant. This model goes by the uninspiring name ΛCDM, where Λ is the Greek symbol used to denote the cosmological constant and CDM stands for Cold Dark Matter. There are models, however, where the dark energy evolves with time; these are often parametrized with the dark energy equation of state w such that

$$\rho_{de}(z) = \rho_{de}(0) \exp\left\{ -3 \int_0^z \frac{dz'}{1+z'} [1 + w(z')] \right\}. \tag{8.10}$$

One goal of cosmology is to determine w; observations are consistent with $w = -1$ at the 20 percent level.

There are a few other parameters that will be encountered as we move through the rest of the chapter, but these are the basic ones that define observables such as distances.

8.2 Statistics of the Density Field

What is meant by the statistics of the density field and why is it important? No theory predicts that a galaxy will form in a given position in the universe or that region X will be under-dense while region Y will be over-dense. Rather, theories make predictions for how matter is distributed on average. The very simplest statistic is the mean of $\delta(\vec{x})$, and we see immediately from the definition in Eq. (8.5) that since the mean value of $\rho(\vec{x})$ is defined to be $\bar{\rho}$, the mean of $\delta(\vec{x})$ vanishes:

$$\langle \delta(\vec{x}) \rangle = 0. \tag{8.11}$$

Here, the angular brackets can be taken to mean the average over all of space.

Consider Fig. 8.2, which shows a slice of an early map of the galaxy distribution in the universe. Even if we force ourselves to ignore the apparent human-like stick figure in the map, there is clear evidence of structure. The simplest way to quantify this would be to take the variance of δ as we did on page 147: $\sigma^2 \equiv \langle \delta^2(\vec{x}) \rangle$. There is a slightly more sophisticated two-point function though, the *correlation function*, that encodes the information about the typical lengths over which matter is clustered. So define

$$\xi(\vec{x}, \vec{y}) \equiv \langle \delta(\vec{x}) \delta(\vec{y}) \rangle \tag{8.12}$$

where again the angular brackets denote an average over all space. Since there is no special place in the universe, we expect the correlation function to depend only on the difference between $\vec{x}$ and $\vec{y}$. That is, *homogeneity* requires

$$\xi(\vec{x}, \vec{y}) = \xi(\vec{x} - \vec{y}). \tag{8.13}$$

We also do not expect any preferred direction, so that the correlations should depend only on the magnitude of the distance $\vec{x} - \vec{y}$, not on its orientation. So *isotropy* requires

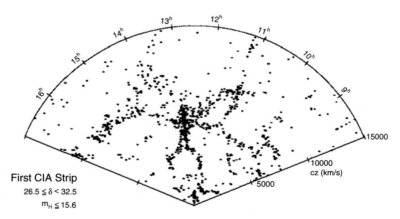

Figure 8.2 Distribution of galaxies in the Center for Astrophysics galaxy survey (de Lapparent et al., 1986). We the observers are situated at the bottom of the image: each point represents a galaxy, so the points at the top are galaxies farthest from us. Labels at the top identify the angular positions (each *hour* corresponds to 15 degrees), while radial labels are in terms of the recession velocities since the more distant galaxies are moving away fastest.

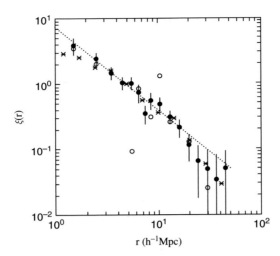

Figure 8.3 The correlation function as measured from galaxies in the QDOT survey (Martinez and Coles, 1994). The different symbols are different ways of estimating the correlation function; they all agree qualitatively that galaxies (and therefore the matter that they trace) are clumped on small scales, where $\xi > 1$, but are distributed relatively smoothly on large scales.

$$\xi(\vec{x} - \vec{y}) = \xi\left(|\vec{x} - \vec{y}|\right). \qquad (8.14)$$

The correlation function of galaxies mapped in the QDOT survey (Lawrence et al., 1999) is shown in Fig. 8.3. A value greater than unity means that two nearby regions are correlated and therefore likely to have similar values of δ. From the figure, we see that this holds for distances smaller than $\sim$ 10 Mpc. On scales larger than this, the correlations are much weaker: if you are sitting in a large over-dense region and look 20 Mpc away, you should not expect to find any over-densities; or, at least, not any more than you would in a random place. This qualitative feature of the correlation function tells us that the universe is clumpy on small scales but smooth on large scales. This fundamental fact about the universe is important and will propagate to the statistics of cosmic shear, so it is worth pointing out that it is obvious: the density of matter in the room in which you are sitting is roughly 1 gm per cm^3, while the mean density in the universe is 10^{-30} gm per cm^3. So on small scales, the universe is very inhomogeneous. It is only by averaging over many small regions, by observing the universe on large scales, that we begin to appreciate its large-scale homogeneity.

The correlation function is intuitive, but its Fourier transform, the *power spectrum*, is perhaps even more useful for comparing with theories and for the calculations that await us. To obtain the power spectrum, we first generalize the expressions in the box on page 155 to three dimensions so that

$$\delta(\vec{x}) = \int \frac{d^3k}{(2\pi)^3} \, \tilde{\delta}(\vec{k}) \, e^{i\vec{k}\cdot\vec{x}}, \tag{8.15}$$

which can be inverted to yield

$$\tilde{\delta}(\vec{k}) = \int d^3x \, \delta(\vec{x}) \, e^{-i\vec{k}\cdot\vec{x}}. \tag{8.16}$$

Again, the mean of $\tilde{\delta}(\vec{k})$ vanishes, but the product of two $\tilde{\delta}$s encodes a lot of information about the structure of the universe. Consider

$$\langle \tilde{\delta}(\vec{k}) \, \tilde{\delta}(\vec{k}') \rangle = \int d^3x \, e^{-i\vec{k}\cdot\vec{x}} \int d^3y \, e^{-i\vec{k}'\cdot\vec{y}} \, \xi(\vec{x} - \vec{y}), \tag{8.17}$$

where the real space $\langle \delta(\vec{x})\delta(\vec{y}) \rangle = \xi(\vec{x} - \vec{y})$ has been used.

We will shortly calculate these integrals and show that they yield the power spectrum, but first let us reconsider the meaning of the angular brackets. So far, we have been thinking of this as an average over all space. We could still think that way, as long as switch from real space to Fourier space, but there is a more profound way to understand this averaging. Our current model of structure in the universe is that, early on, every Fourier mode $\tilde{\delta}(\vec{k})$ was a random variable drawn from a Gaussian distribution. Therefore, the angular brackets are perhaps best understood as the mean value of the quantity drawn from this distribution. For example, if the distribution were still Gaussian, then it would be proportional to $\exp\{-|\tilde{\delta}(\vec{k})|^2/2P(k)\}$, where $P(k)$ is the power spectrum. Then, the expected value of $|\tilde{\delta}(\vec{k})|^2$ would be proportional to $P(k)$. Today, due to the long-term effects of gravity, the distribution is no longer Gaussian, but there is still some underlying distribution from which the $\tilde{\delta}(\vec{k})$ are drawn. The angular brackets in Eq. (8.17) are best thought of then as referring to the expected value of $\tilde{\delta}(\vec{k}) \, \tilde{\delta}(\vec{k}')$ when drawn from the distribution.

Back to the calculation in Eq. (8.17), replacing the inner dummy variable with $\vec{x}_- = \vec{x} - \vec{y}$ leads to

$$\langle \tilde{\delta}(\vec{k}) \, \tilde{\delta}(\vec{k}') \rangle = \int d^3x \, e^{-i(\vec{k}+\vec{k}')\cdot\vec{x}} \int d^3x_- e^{i\vec{k}'\cdot\vec{x}_-} \, \xi(\vec{x}_-). \tag{8.18}$$

Then, switching the order of integration leads to a three-dimensional Dirac delta function:

$$\langle \tilde{\delta}(\vec{k}) \, \tilde{\delta}(\vec{k}') \rangle = (2\pi)^3 \, \delta_D^3(\vec{k} + \vec{k}') \int d^3x_- e^{i\vec{k}'\cdot\vec{x}_-} \, \xi(\vec{x}_-)$$

$$= (2\pi)^3 \, \delta_D^3(\vec{k} + \vec{k}') \, P(k). \tag{8.19}$$

The term multiplying the Dirac delta function is the Fourier transform of the correlation function, called the *power spectrum*, $P(k)$. Note that, since ξ depends only on the magnitude of $\vec{x}_-$, the power spectrum depends only on the magnitude of its argument $P = P(|\vec{k}|)$. Note also that the Dirac delta function enforces the fact that fluctuations in the Fourier modes are correlated with one another only if the two modes have equal and opposite wave vectors. Otherwise, the fluctuations are uncorrelated: that is, the average value of $\tilde{\delta}(\vec{k})\tilde{\delta}(\vec{q})$, when $q \gg k$ for example vanishes. This property of the density distribution follows directly above from the fact that the correlation function depends only on the distance between two points; that is, it follows from the fact that the universe is homogeneous, with no special location.

Figure 8.4 shows the power spectrum as measured from the positions of galaxies in the Baryon Oscillation Spectroscopic Survey (BOSS). Note the much wider range of scales than is contained in Fig. 8.3. There, the largest scales probed corresponded to distances of order 30–40 Mpc. The largest scales probed by BOSS, a survey concluded almost twenty years after QDOT, are represented in the figure by the lowest k-modes, with $r \simeq \pi/k$. So the largest scales probed by BOSS are of order 300 Mpc, an order of magnitude larger than the earlier survey.

The plot of the power spectrum in the form of Fig. 8.4 hides some of the features that are more evident in the correlation function. In particular, it looks as though there is *more* power on large scales than on small scales, and we have come to expect the opposite: a universe smooth on large scales but clumpy on small. To understand that this results from a misinterpretation of the figure, it pays to reconsider the simplest measure of the fluctuations in the density field, the variance

$$\sigma^2 \equiv \langle \delta^2(\vec{x}) \rangle. \tag{8.20}$$

Fourier transforming both $\delta(\vec{x})$s leads to

$$\sigma^2 = \int \frac{d^3k}{(2\pi)^3} \int \frac{d^3k}{(2\pi)^3} e^{i\vec{x}\cdot(\vec{k}+\vec{k}')} \langle \tilde{\delta}(\vec{k})\,\tilde{\delta}(\vec{k}') \rangle, \tag{8.21}$$

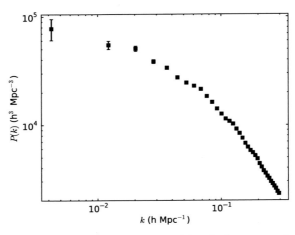

Figure 8.4 The power spectrum, as measured by Anderson et al. (2014) using data from the Baryon Oscillation Spectroscopic Survey (BOSS).

which simplifies with the aid of Eq. (8.19) to

$$\sigma^2 = \int \frac{d^3 k}{(2\pi)^3} P(k)$$
$$= \int_0^\infty \frac{dk}{k} \frac{k^3 P(k)}{2\pi^2}. \tag{8.22}$$

Think of this integral as the sum over logarithmic bins in k: $d\ln(k) = dk/k$. In each bin, the contribution to the variance, to the fluctuations, is given by the *dimensionless* combination $k^3 P(k)/2\pi^2$. When this number is small, the fluctuations on the scale k are small. And since the combination is dimensionless, *small* means something concrete: as long as $k^3 P(k)/2\pi^2$ is less than one, the fluctuations are linear, so that an observer smoothing on these scales would observe a nearly homogeneous universe. When it is larger than unity, then the fluctuations have gone nonlinear, and an observer might see a void or a large over-density. With this in mind, consider Fig. 8.5, which contains the same data as was plotted in Fig. 8.4, simply multiplied with k^3 to produce the appropriate dimensionless measure of clustering. We see again that on large scales, this measure is small, less than unity, indicating that the universe is smooth on large scales. On small scales $k > 0.1h$ Mpc^{-1}, this measure rises above unity, a sign that the universe is very clumpy on small scales.

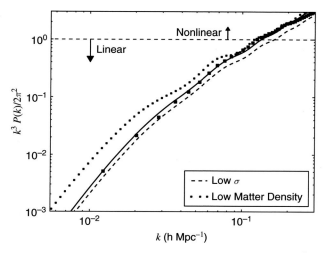

Figure 8.5 Same data as in Fig. 8.4, but here $P(k)$ is multiplied by k^3. The dimensionless combination is less than one on large scales where the universe is smooth, and larger than one on small scales, where the density has gone nonlinear. Dashed horizontal line delineates the region between the scales that have gone nonlinear and those that have not. Solid, dashed, and dotted curves are the predictions for the power spectrum with three different sets of cosmological parameters

The first lesson is that the power spectrum quantifies fluctuations in Fourier space as a function of scale. We will soon generalize this to the 2D power spectrum of shear fluctuations and will see that $l^2 C_l / 2\pi$, where C_l is the 2D power spectrum, serves the same function when it comes to quantifying the fluctuations in a field that varies in two dimensions, for example across the sky. The second point evident in Fig. 8.5 is that the power spectrum depends on the cosmological model. The figure shows three different models, each of which is specified by a different set of parameters. The model depicted by the dashed curve has a lower value of σ than does the fiducial solid curve, while the dotted curve shows a model with the same value of σ but a lower density of matter than the fiducial model. The take-away is that, by measuring the power spectrum, we can infer the values of the cosmological parameters.

8.3 Cosmic Shear Decomposition

Just as the three-dimensional density field is inhomogeneous across the sky encoded in the fractional over-density $\delta(\vec{x})$, the two-dimensional cosmic shear field varies as a function of angle $\vec{\theta}$. There is a tight connection between $\delta(\vec{x})$ and the shear fields because the shear along a given line of sight is produced by the over-density in that direction. However, besides the dimensionality, another difference between the fields is that the shear field has two components $\gamma_1(\vec{\theta})$ and $\gamma_2(\vec{\theta})$; let's explore the implications of these two components first and then return to the statistics of cosmic shear and how they are related to the statistics of the over-density field.

Consider again the two components of shear in Fourier space, as given in Eq. (8.3). Define the angle ϕ_l to be that which the vector $\vec{l}$ makes with respect to the x-axis (which can be chosen arbitrarily and then fixed). Then, the x-component of $\vec{l}$, $l_x = l \cos \phi_l$, and more generally we can rewrite the two Fourier components as

$$\tilde{\gamma}_1(\vec{l}) = -\frac{l^2 \tilde{\Phi}(\vec{l})}{2c^2} \left[\cos^2 \phi_l - \sin^2 \phi_l \right]$$

$$\tilde{\gamma}_2(\vec{l}) = -\frac{l^2 \tilde{\Phi}(\vec{l})}{2c^2} \left[2 \cos \phi_l \sin \phi_l \right]. \tag{8.23}$$

Using familiar trigonometric identities leads to even simpler expressions:

$$\tilde{\gamma}_1(\vec{l}) = -\frac{l^2 \tilde{\Phi}(\vec{l})}{2c^2} \cos(2\phi_l)$$

$$\tilde{\gamma}_2(\vec{l}) = -\frac{l^2 \tilde{\Phi}(\vec{l})}{2c^2} \sin(2\phi_l). \tag{8.24}$$

Note the similarity with the angular dependence in the case of shear around a spherically symmetric object, as given in Eqs. (7.4) and (7.5). We have had to work a little harder here, by going into Fourier space, but clearly there is a tight connection between the very simple shear pattern around a spherically symmetric object and the pattern induced by a general inhomogeneity.

Just as in the case of a spherically symmetric lens, we identified a linear combination of the two components of shear that contained the signal (the tangential shear) and another linear combination that should vanish, we can do the same in the cosmic shear case. Consider

$$\tilde{E}(\vec{l}) \equiv -\tilde{\gamma}_1(\vec{l}) \, \cos(2\phi_l) - \tilde{\gamma}_2(\vec{l}) \, \sin(2\phi_l)$$
$$= \frac{-l^2 \tilde{\Phi}(\vec{l})}{2c^2}. \tag{8.25}$$

Then, $\tilde{E}$ should be equal to $\tilde{\kappa}$, and therefore contain the signal of interest, with several caveats. First, both components of shear contain noise and these will also be present in $\tilde{E}$, so $\tilde{E}$ can be thought of as a linear combination of the data that contains the signal ($\tilde{\kappa}$) and noise. On a more subtle level, the apparent equality between $\tilde{E}$ and $\tilde{\kappa}$ rests on the implicit assumption that the only perturbation causing deflection is the gravitational potential. This has been our assumption throughout and it has served us well, so we will not abandon it now. But on a purely formal level, it is worth noting that the metric could contain other perturbations – for example, gravitational waves – and these would alter the relationship between $\tilde{E}$ and $\tilde{\kappa}$.

The connection between the E-mode defined here and the tangential shear around a single object propagates to mass maps. Indeed, Eq. (7.52) is identical to Eq. (8.25) here and this raises the interesting possibility of constructing mass maps not only when the distortions are due to a single object, as in Fig. 7.13, but also when they are due to all structure along the line of sight. Fig. 8.6 shows such a map, constructed in a way very similar to the technique that produced Fig. 7.13.

This κ map is due to all structure along the line of sight, so it is useful to combine Eqs. (8.1) and (8.2) to obtain

$$\kappa(\vec{\theta}) = \frac{1}{c^2 D_S} \nabla_\theta^2 \int_0^{D_S} dD_L \frac{D_{SL}}{D_L} \phi(x^i = D_L \theta^i, D_L)$$
$$= \frac{1}{c^2} \int_0^{D_S} dD_L \frac{D_{SL} D_L}{D_S} \nabla^2 \phi(x^i = D_L \theta^i, D_L) \tag{8.26}$$

where the Laplacian with respect to the angle θ has been replaced by the ordinary Laplacian with respect to position $\vec{x}$: $\nabla_\theta^2 \rightarrow D_L^2 \nabla^2$ since ϕ depends on $\vec{\theta}$ only through the combination $D_L \vec{\theta}$. Using Poisson's equation now leads to a new interpretation of κ:

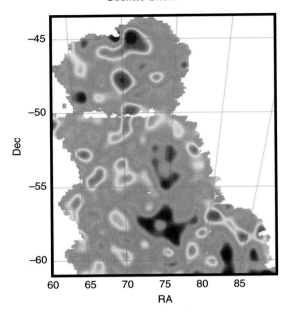

Figure 8.6 Map of convergence κ over 150 square degrees of early data from the Dark Energy Survey (Vikram et al., 2015) constructed from shapes of over a million background galaxies. Blue regions are underdense and red overdense, with the extreme values of κ reaching to ± 0.015. This map includes contributions from all matter along the line of sight to the background galaxies, which are distributed around $z \simeq 0.8$. (See color plates section.)

$$\kappa(\vec{\theta}) = \int_0^{D_S} dD_L \, W(D_L, D_S) \, \delta(D_L \vec{\theta}, D_L) \tag{8.27}$$

where the function that weights the density along the line of sight is defined as

$$W(D_L, D_S) \equiv \frac{D_{SL} D_L}{D_S} \frac{4\pi G}{c^4} \bar{\rho}_m(a) a^2(D_L). \tag{8.28}$$

So κ is a measure of the over-density field integrated along the line of sight with weighting function W. Note that the scale factor here is a function of the dummy variable D_L in the sense that it is evaluated at the epoch when light from the background galaxy is a distance D_L from us. We can simplify a bit by noting that the matter density becomes dilute as the universe expands, falling as the volume, so

$$\bar{\rho}_m(a) = \bar{\rho}_m(t_0) \, a^{-3}$$
$$= \Omega_m \rho_{cr} \, a^{-3} \tag{8.29}$$

where $\bar{\rho}_m(t_0)$ is the present value of the matter density. We have expressed this in terms of the constant critical density of the universe in order to utilize the

Friedmann equation (8.7) to arrive at a more compact version of the weighting function

$$W(D_L, D_S) \equiv \frac{3\Omega_m H_0^2}{2c^2 a(D_L)} \frac{D_{SL} D_L}{D_S}. \tag{8.30}$$

The E-mode identified in Eq. (8.25) is a linear combination that contains the information of interest. Since there are two components of shear, there is another linear combination that also proves useful, although for a different reason. Consider

$$\tilde{B}(\vec{l}) \equiv \tilde{\gamma}_1(\vec{l}) \sin(2\phi_l) - \tilde{\gamma}_2(\vec{l}) \cos(2\phi_l). \tag{8.31}$$

Given the dependence on $\tilde{\Phi}$ shown in Eq. (8.24), this so-called B-mode should vanish. As such, it is an excellent test of systematic errors. Any survey must show not only a nonzero E-mode, which contains the signal, but also a B-mode consistent with zero to convince itself and the broader community that its measurements are reliable.

8.4 Statistics of Convergence

Consider the mass map in Fig. 8.6. We now know that it represents the over- and under-density of matter in the universe integrated all the way back to the source galaxies whose shapes have been measured. In that sense, it shares features with the 3D map of the galaxies in the universe shown in Fig. 8.2: there are under- and over-dense regions and the mean over-density in each case $\langle \delta \rangle = \langle \kappa \rangle = 0$. Just as in the 3D case, we argued that the information resides in the statistical properties of the maps beyond the mean, for κ as well we want to examine its statistical properties. And just as in the 3D case, the most powerful statistic is the two-point function in the form of the real space correlation function or its cousin the Fourier space power spectrum.

We will focus on the latter here, so let us first write down the definition of the power spectrum in two dimensions, explicitly following the 3D definition/result of Eq. (8.19):

$$\langle \tilde{\kappa}(\vec{l}) \tilde{\kappa}(\vec{l}') \rangle = (2\pi)^2 \delta_D^2(\vec{l} + \vec{l}') C_l^{\kappa\kappa}. \tag{8.32}$$

Just as in the three-dimensional case, the power spectrum depends only on the magnitude of the wave vector $\vec{l}$, not its direction. The analogy with the 3D case means that each angular scale l (and here, too, large l corresponds to small scales and vice versa) has fluctuations with variance $l^2 C_l^{\kappa\kappa}/2\pi$. The variance of the fluctuations (Exercise (8.5)) of the convergence field will be the logarithmic integral over all l of $l^2 C_l^{\kappa\kappa}/2\pi$.

Before diving into the calculation of the $C_l^{\kappa\kappa}$s, let us first get a picture of what these Fourier modes look like. Fig. 8.7 shows two such modes, one with small l

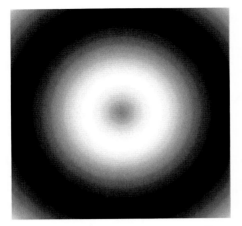

Figure 8.7 The convergence $\kappa(\vec{\theta})$ from a single Fourier mode, each with $l_x = l_y = l/\sqrt{2}$. Left panel has low l so varies only on large angular scales; right panel has much larger l so varies over small angular scales.

varying over large angular scales only and the other with much larger l, so $\kappa(\vec{\theta})$ changes sign many times over the same angular region in the sky. A bit more quantitatively, a single mode leads to angular variation that scales as $\sin(l\theta)$, so the subsequent peaks in κ shown in Fig. 8.7 occur at $\theta = 0, \pi/l, 2\pi/l, \dots$ In other words, a Fourier mode with wavenumber l induces variations on an angular scale $\theta \sim \pi/l$.

With this picture in mind, let us first obtain an analytic expression for $\tilde{\kappa}(\vec{l})$ by Fourier transforming Eq. (8.27):

$$\tilde{\kappa}(\vec{l}) = \int d^2\theta\, e^{-i\vec{l}\cdot\vec{\theta}} \int_0^{D_S} dD_L\, W(D_L, D_S)\, \delta(D_L\vec{\theta}, D_L). \qquad (8.33)$$

We want to carry out the $d^2\theta$ integral, so it makes sense to Fourier transform the three-dimensional over-density δ; in that way, its $\vec{\theta}$ dependence becomes trivial:

$$\tilde{\kappa}(\vec{l}) = \int d^2\theta\, e^{-i\vec{l}\cdot\vec{\theta}} \int_0^{D_S} dD_L\, W(D_L, D_S) \int \frac{d^3k}{(2\pi)^3}\, e^{i\vec{k}\cdot[D_L\vec{\theta}, D_L]}\, \tilde{\delta}(\vec{k}). \qquad (8.34)$$

The second exponential now contains the dot product between the 3D wave vector $\vec{k}$ and the 3D position, composed of a transverse part $D_L\vec{\theta}$ and a radial part D_L. It pays then to decompose the wave vector as well into a transverse part $k_\perp$ and a part parallel to the line of sight k_z. A Fourier mode that varies along the line of sight will have large k_z while one that is constant in the direction towards the source but varying in the transverse plane will have $k_\perp \gg k_z$. In any event, the $d^2\theta$ integral now yields a Dirac delta function that enforces the equality $\vec{l} = D_L\vec{k}_\perp$, so we have

$$\tilde{\kappa}(\vec{l}) = \int_0^{D_S} dD_L\, W(D_L, D_S) \int \frac{dk_z}{2\pi} e^{ik_z D_L} \int d^2k_\perp\, \delta_D^2(\vec{l} - D_L\vec{k}_\perp)\, \tilde{\delta}(\vec{k}_\perp, k_z).$$
(8.35)

We can use the Dirac delta function to perform the $d^2k_\perp$ integral, as long as we first use Eq. (2.74) to write $\delta_D^2(\vec{l} - D_L\vec{k}_\perp) = \delta_D^2(\vec{k}_\perp - \vec{l}/D_L)/D_L^2$. Then, our final expression for the Fourier transform of the convergence becomes

$$\tilde{\kappa}(\vec{l}) = \int_0^{D_S} dD_L\, \frac{W(D_L, D_S)}{D_L^2} \int \frac{dk_z}{2\pi} e^{ik_z D_L}\, \tilde{\delta}(\vec{l}/D_L, k_z).$$
(8.36)

Intuition in Fourier space takes time to build up, but here we can see two interesting features of $\tilde{\kappa}(\vec{l})$. First, any perturbations that vary rapidly along the line of sight (quantitatively, $k_z D_L \gg 1$) will not contribute to the convergence for the oscillating exponential leads to cancellations. A 3D Fourier mode that varies only in the radial direction is depicted in the left panel of Fig. 8.8, and it is clear that κ will be zero from this mode because of the cancellations between under- and over-dense regions.

A second feature captured in Eq. (8.36) is that, for a fixed angular scale, the variation in κ comes from 3D density Fourier modes with wavenumber $k = k_\perp =$

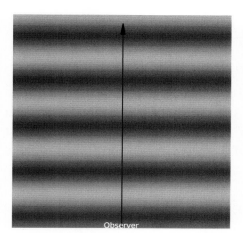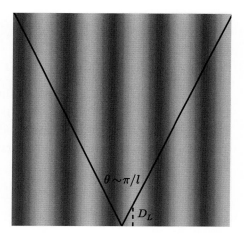

Figure 8.8 The impact of 3D density perturbations on $\kappa(\theta)$, which is an integral along the line of sight. Over-dense regions are shaded white and under-dense dark, and we the observer are situated at the bottom center measuring variations over angular scales $\theta \sim \pi/l$. *Left panel:* This long wavelength mode has large k_z and $k_\perp = 0$; its variations along the line of sight lead to cancellation so κ will get no contributions from this mode, or any mode with large k_z. *Right panel:* Mode with $k_z = 0$ but nonzero $k_\perp$. Along the line of sight, the only regions that contribute to $\tilde{\kappa}(l)$ are those for which D_L is small; otherwise the peaks and troughs cancel. The maximum contribution from this Fourier mode to the difference between κ along the two lines of sight separated by an angle π/l comes from a distance D_L away. Note that the wavelength of this Fourier mode is $\lambda \sim D_L\theta$, or $k_\perp \sim l/D_L$.

l/D_L. These have wavelength $\pi/k = \pi D_L/l$. Since the scale l is associated with variations on angular scales $\theta \sim \pi/l$, this means that the angular scales that contribute to $\kappa(\vec{l})$ are of order $\theta \sim \pi/l$ and these pick up contributions from 3D perturbations with wavelengths $\lambda \sim D_L\theta$. This can be gleaned from the right panel of Fig. 8.8. For fixed angular scale θ, the perturbations far from us do not contribute to κ because of the cancellations. So the main contribution comes from the region a distance $D_L \sim \lambda/\theta$ where λ is the wavelength of the perturbation (of order $\pi/k_\perp$). Therefore, the dominant contribution to $\tilde{\kappa}(l)$ comes from distance and wavenumbers that satisfy $k_\perp D_L \sim l$.

There is a time for trying to understand and a time for calculating. Let's begin the calculation of the power spectrum of the convergence. A simple starting point is to integrate Eq. (8.32) over all $\vec{l}'$ so that, with the aid of the Dirac delta function, the right-hand side is simply equal to $C_l^{\kappa\kappa}$, modulo factors of 2π. Inserting the expression in Eq. (8.36) leads to

$$
\begin{aligned}
C_l^{\kappa\kappa} &= \int \frac{d^2l'}{(2\pi)^2} \langle \tilde{\kappa}(\vec{l})\tilde{\kappa}(\vec{l}')\rangle \\
&= \int \frac{d^2l'}{(2\pi)^2} \int_0^{D_S} dD_L \frac{W(D_L, D_S)}{D_L^2} \int \frac{dk_z}{2\pi} e^{ik_z D_L} \\
&\quad \times \int_0^{D_S} dD_L' \frac{W(D_L', D_S)}{D_L'^2} \int \frac{dk_z'}{2\pi} e^{ik_z' D_L'} \\
&\quad \times \langle \tilde{\delta}(\vec{l}/D_L, k_z)\tilde{\delta}(\vec{l}'/D_L', k_z')\rangle.
\end{aligned}
\tag{8.37}
$$

The last line here is the simplest: it is just Eq. (8.19) with a particular choice of $\vec{k}$ and $\vec{k}'$, so it is equal to a product of the 3D matter power spectrum $P(k)$[1] and a Dirac delta function that equates k_z and k_z' and also a 2D Dirac delta function that equates $\vec{l}/D_L$ and $\vec{l}'/D_L'$. Using the first of these to eliminate the k_z' integral leads to

$$
\begin{aligned}
C_l^{\kappa\kappa} &= \int \frac{d^2l'}{(2\pi)^2} \int_0^{D_S} dD_L \frac{W(D_L, D_S)}{D_L^2} \int_0^{D_S} dD_L' \frac{W(D_L', D_S)}{D_L'^2} \\
&\quad \times \int \frac{dk_z}{2\pi} P(\vec{l}/D_L, k_z) e^{ik_z(D_L - D_L')} (2\pi)^2 \delta_D^2 \left(\frac{\vec{l}}{D_L} + \frac{\vec{l}'}{D_L'}\right).
\end{aligned}
\tag{8.38}
$$

Let's use the 2D Dirac delta function to perform the d^2l' integral, remembering to multiply by $D_L'^2$ as in Eq. (2.74). Then,

$$
C_l^{\kappa\kappa} = \int_0^{D_S} dD_L \frac{W(D_L, D_S)}{D_L^2} \int_0^{D_S} dD_L' \, W(D_L', D_S)
$$

---

[1] Although any field has a power spectrum associated with it, in cosmology the 3D power spectrum of the fractional overdensity has a special status so is often referred to simply as the *power spectrum* or sometimes the *matter power spectrum*.

$$\times \int \frac{dk_z}{2\pi} P(\vec{l}/D_L, k_z) \, e^{ik_z(D_L - D'_L)}. \tag{8.39}$$

The argument of the power spectrum is $k = \sqrt{(l/D_L)^2 + k_z^2}$. But we saw from Fig. 8.8 that the only modes that contribute are those with $k_z D_L \lesssim 1$. Therefore, for large l, $l/D_L \gg k_z$, and we can safely set the argument of the power spectrum to $k = l/D_L$. Again this conforms to our intuition that it is only the transverse modes that will be contributing to the convergence, which – since it is an integral over the line of sight – is insensitive to radial variations. This approximation is crucial, for now the k_z integral simply yields a Dirac delta function in $D_L - D'_L$; after performing the D'_L integral, we are left with

$$C_l^{\kappa\kappa} = \int_0^{D_S} dD_L \, \frac{W^2(D_L, D_S)}{D_L^2} P(l/D_L). \tag{8.40}$$

We have been a little sloppy with one aspect of this: the time-dependence of the 3D power spectrum. Recall that the over-density δ, whose square leads to the power spectrum, is to be evaluated at times along the line of sight. Therefore, the power spectrum too should be evaluated at the epoch corresponding to the time when light from the source galaxies at D_S are a distance D_L from us: $P \to P(l/D_L; a(D_L))$. You can tabulate the values of a or z that correspond to a given D_L in Exercise (8.6).

One final tweak to Eq. (8.40): the weighting function W – sometimes called the *window function* – implicitly assumes that all the background galaxies are at a single redshift. In any realistic situation, the background galaxies will be spread over a finite extent in redshift, and the window function must account for this. This is typically quantified by introducing the redshift distribution of the background galaxies, dn/dz. The generalization of Eq. (8.30) is then

$$W(D_L) = \frac{3\Omega_m H_0^2}{2c^2 a(D_L)} \int_0^{\infty} dz \, \frac{dn}{dz} \frac{D_{SL}(z) D_L}{D_S(z)} \, \Theta(D_S(z) - D_L) \tag{8.41}$$

where the Heavyside step function Θ is equal to one if its argument is positive and zero otherwise. It enforces the constraint that the sources must be behind the lenses. Fig. 8.9 shows an example of the background galaxy distribution in two redshift bins (these are actual bins used in the Canada–France–Hawaii Telescope (CFHT) lensing project (Kilbinger et al., 2013)). Note that the redshift distribution is normalized so that the integral over all redshifts is equal to one (for each bin). Also, note that the predicted spectrum is larger for galaxies in the higher redshift bin, as the light from them propagates through more of the universe. But the larger point is that by measuring the cosmic shear spectrum in multiple redshift bins, we can – via Eq. (8.40) – obtain an estimate of the power spectrum and its evolution.

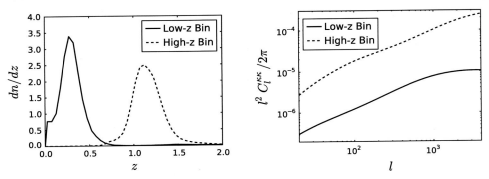

Figure 8.9 Sample distribution of background galaxies and the predicted convergence spectra for those two bins using Eq. (8.40). Note that the convergence from galaxies further away is larger since the light they emit passes through more of the inhomogeneous universe.

8.5 Errors on the Convergence Power Spectrum

We mentioned after Eq. (8.25) that the E-mode will contain both signal and noise. We have computed the signal; it is time to compute the noise. It turns out there are (at least) two sources of noise when estimating $C_l^{\kappa\kappa}$; the first is shape noise, the second, more esoteric, is cosmic variance.

Let's examine the impact of shape noise on the $C_l^{\kappa\kappa}$ first. Suppose there was no signal at all. We have shown in §7.2 that the error on the shear in a given region will be $\sigma_{\rm sh}/\sqrt{N}$ where N is the number of galaxies used to estimate the shear. Quantitatively, we write that the estimator for shear has mean zero and variance

$$\langle \hat{\gamma}_{\alpha,i} \hat{\gamma}_{\beta,j} \rangle = \delta_{\alpha\beta}\, \delta_{ij}\, \frac{\sigma_{\rm sh}^2}{N} \tag{8.42}$$

where α, β label the component of shear (either 1 or 2) and i, j label pixels with N galaxies in them. How does this error on the shear in a pixel propagate to the error on $C_l^{\kappa\kappa}$? Returning to Eq. (8.25), we see that $\tilde{E}(\vec{l})$ is a linear combination of the Fourier transforms of the two components of the shear. Nominally, then, we would write:

$$\tilde{E}(\vec{l}) = \int d^2\theta\, e^{-i\vec{l}\cdot\vec{\theta}} \left[\gamma_1(\vec{\theta})\, \cos(2\phi_l) + \gamma_2(\vec{\theta})\, \sin(2\phi_l) \right] \tag{8.43}$$

with ϕ_l the angle between the vector $\vec{l}$ and an arbitrary x-axis. To make contact with our expression for shape noise, though, which Eq. (8.42) expresses in terms of real space pixels, let us discretize these Fourier transforms so that

$$\tilde{E}(\vec{l}) = (\Delta\theta)^2 \sum_i e^{-i\vec{l}\cdot\vec{\theta}_i} \left[\gamma_{1,i}\, \cos(2\phi_l) + \gamma_{2,i}\, \sin(2\phi_l) \right]. \tag{8.44}$$

Here, the integral has been changed to a sum over pixels i, each with area $(\Delta\theta)^2$. The mean of this, of course, is zero, but the variance is not:

$$\langle \tilde{E}(\vec{l})\tilde{E}(\vec{l}')\rangle = (\Delta\theta)^4 \frac{\sigma_{sh}^2}{N} \sum_i e^{-i\vec{\theta}_i\cdot(\vec{l}+\vec{l}')}$$

$$\times \left[\cos(2\phi_l)\cos(2\phi_{l'}) + \sin(2\phi_l)\sin(2\phi_{l'})\right]. \qquad (8.45)$$

To carry out the sum over i, transform it back into an integral: $(\Delta\theta)^2 \sum_i \rightarrow \int d^2\theta$. Then, the integral over $d^2\theta$ is $(2\pi)^2\delta^2(\vec{l}+\vec{l}')$. The delta function ensures that the cosines and sines sum to one, and then picking out the coefficient of $(2\pi)^2\delta^2(\vec{l}+\vec{l}')$ as the contribution of the noise to the spectrum:

$$C_l^n = \frac{(\Delta\theta)^2\sigma_{sh}^2}{N} = \frac{\sigma_{sh}^2}{n} \qquad (8.46)$$

where n is the galaxy density $N/(\Delta\theta)^2$. This noise contribution must be subtracted from the measured spectrum to get an estimate of the contribution from the signal, $C_l^{\kappa\kappa}$. Note that while the signal has the shape depicted in Fig. 8.9, the noise is flat, so $l^2 C_l^n$ scales as l^2. Therefore, shape noise will eventually overtake the signal on small scales.

To understand the other source of noise, cosmic variance, it is useful to think about how one would go about estimating $C_l^{\kappa\kappa}$ practically. After obtaining the shape estimates of all background galaxies, we could transform this *shear cata-log* into a map of shear by pixelizing and then averaging the shapes of all galaxies in a single pixel to obtain $\gamma_{1,i}$ and $\gamma_{2,i}$. With a large enough contiguous region, we could then take the Fourier transform of this pixelized map of $\gamma_{1,i}$ and $\gamma_{2,i}$ and estimate $\tilde{E}(\vec{l})$ via Eq. (8.44). To estimate the spectrum, the next step would be to identify all modes with a given amplitude $|\vec{l}| = l$ (see Fig. 8.10 for a simple example). Then, an estimate of the spectrum C_l would be the mean value of $|\tilde{E}(\vec{l})|^2$ of all the selected modes.

This means that we are using a finite number of modes to estimate $C_l^{\kappa\kappa}$. Fig. 8.11 shows an example of this: a histogram of values of $|\tilde{E}(\vec{l})|$ from a finite number of modes. This histogram depicts a very profound aspect of cosmology, perhaps the most difficult part of cosmology to understand. We are trying to learn about the distribution of matter in the universe, in this case by studying the shear field. To do this, we sample the field at many different places; each sample by itself teaches us little, but by studying the distribution of samples, for example in the histogram of $|\tilde{E}(\vec{l})|$, we can learn about the underlying distribution from which the samples are drawn. In the figure, there are 100 different modes whose amplitude $|\tilde{E}(\vec{l})|$ is measured. We are interested in the spread of these in order to determine the true value of $C_l^{\kappa\kappa}$, which is the variance of $|\tilde{E}(\vec{l})|$. Clearly, the more samples we have,

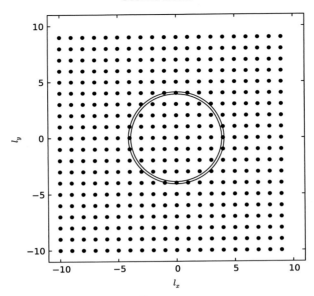

Figure 8.10 Dots denote 2D vectors $\vec{l} = (l_x, l_y)$ in discrete Fourier space. All the dots within the two circles have $|\vec{l}|$ close to $l = 4$ so would be averaged over to produce an estimate of C_4. The spacing between dots is determined by the area of the survey $A = \Theta^2$, with, e.g., $l_x = \ldots, \frac{1}{2\pi\Theta}, \frac{2}{2\pi\Theta}, \frac{3}{2\pi\Theta}, \ldots$

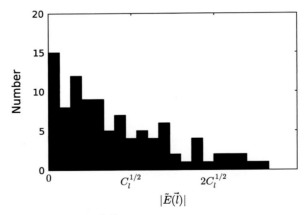

Figure 8.11 A histogram of $|\tilde{E}(\vec{l})|$ used to estimate $C_l^{\kappa\kappa}$. The width of the distribution should be equal to C_l, but for a finite number of $\vec{l}$s, there is always difficulty estimating the true value of the width. In the example shown here, $N_l = 100$ and the estimated value of the spectrum is $1.09 C_l$.

the better we will do at estimating the width of the distribution, in obtaining the $C_l^{\kappa\kappa}$s. You can work through the algebra for the case of a Gaussian distribution in Exercises (8.9) and (8.10); the result is that the uncertainty on the spectrum due to the finite number of modes is

$$\Delta C_l^{KK} = \left(\frac{2}{N_l}\right)^{1/2} C_l^{KK} \tag{8.47}$$

where N_l is the number of l-modes used in the estimate. This inevitable error due to finite number of modes used in the estimate is called *cosmic variance*.

There are two further tweaks to the formula for the error on the spectrum in Eq. (8.47). The first is simple: the spectrum gets contributions from both signal and shape noise, so we can include both of those, leading to:

$$\Delta C_l^{KK} = \left(\frac{2}{N_l}\right)^{1/2} [C_l^{KK} + C_l^n]. \tag{8.48}$$

The second tweak is to calculate N_l, the number of modes that will contribute to an estimate of C_l^{KK} for a given l. Looking back at Fig. 8.10, we see that the number of modes that will contribute is equal to the density of states multiplied by the area of the annulus within which l_x, l_y must lie. The latter is simply $2\pi l \Delta l = 2\pi l$ if we focus on a single l (so that $\Delta l = 1$). To get the density of states, first consider 1D Fourier space, where the size of the region in real space is Θ. Then, each Fourier mode has value: $2\pi n/\Theta$, where n is an integer. The number of modes goes up by one as l changes by $2\pi/\Theta$. Therefore, $\Delta N_l/\Delta l = 1/(2\pi/\Theta)$, and this is the density of states in one dimension. Squaring this to get the two-dimensional density and multiplying by the area yields

$$N_l = (2\pi l) \times \left(\frac{\Theta}{2\pi}\right)^2$$
$$= \frac{A}{4\pi} 2l \tag{8.49}$$

where $A \equiv \Theta^2$ is the area of the survey. Since the total available area on the sky is 4π, the first ratio is simply the fraction of the sky covered by the survey; let's call it f_{sky}. We arrive at our final result for the uncertainty on the spectrum from a survey that covers f_{sky} and contains n galaxies per radian squared:

$$\Delta C_l^{KK} = \left(\frac{1}{f_{sky}l}\right)^{1/2} \left[C_l^{KK} + \frac{\sigma_{sh}^2}{n}\right]. \tag{8.50}$$

An interesting dividing point is the l above which shape noise dominates. Since $l^2 C_l^{KK}/2\pi$ is roughly constant, the first term in square brackets in Eq. (8.50) falls off as l^{-2} while the second is constant, so the point at which shape noise begins to dominate is

$$l_m \equiv \left[\frac{l^2 C_l^{KK}}{2\pi} \frac{2\pi n}{\sigma_{sh}^2}\right]^{1/2} \tag{8.51}$$

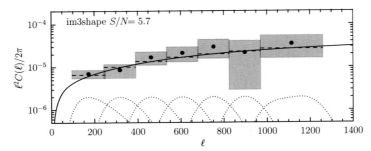

Figure 8.12 Spectrum of the convergence from early data taken by the Dark Energy Survey (Becker et al., 2016). The solid curve is the best fit model in the context of ΛCDM. The widths of the rectangles denote the l-range over which each measurement is averaged, so $\Delta l \sim 150$; the heights of the rectangles are the error bars; im3shape refers to the algorithm used to measure the shapes of the background galaxies (a newer version of im2shape described on page 125). This dataset covered 139 square degrees, corresponding to $f_{sky} = 0.0034$, with a galaxy density $n = 4$ per square arcminute.

Setting $\sigma_{sh} \rightarrow 0.2$ leads to an estimate of

$$l_m \sim 1400 \left(\frac{n}{10 \, \text{arcminute}^{-2}} \right)^{1/2} \left(\frac{l^2 C_l^{\kappa\kappa}/2\pi}{10^{-4}} \right)^{1/2}. \tag{8.52}$$

Fig. 8.12 shows an example, which will allow us to test some of these formulae. The data come from 139 square degrees with a galaxy density of 4 per square arcminute. These features mean that the first point at $l \simeq 175$ is dominated by cosmic variance, while the last, at $l \simeq 1100$, is dominated by shot noise. You can estimate the corresponding error bars using Eq. (8.50) in Exercise (8.11) and show that they are comparable to those shown in Fig. 8.12.

8.6 Intrinsic Alignment

We have implicitly assumed until now that, apart from the distortion caused by cosmic shear, the ellipticity of a given galaxy is completely independent of its neighbors' ellipticities. This is not necessarily true. Ellipticities are partially determined by the local gravitational field, so if several galaxies reside in the same field, their ellipticities will be at least somewhat correlated. Fig. 8.13 shows that if there is a gradient in the gravitational field, a galaxy can be stretched in one direction or another.

Consider this effect in the case depicted in the figure where the field varies only in one direction. Then, Taylor expanding the gravitational potential in the vicinity of a massive object like a galaxy,

$$\phi(x) \simeq \phi(0) + \left. \frac{\partial \phi}{\partial x} \right|_{x=0} x + \frac{1}{2} \left. \frac{\partial^2 \phi}{\partial x^2} \right|_{x=0} x^2. \tag{8.53}$$

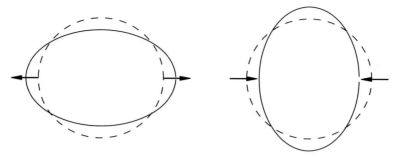

Figure 8.13 Dashed circles show the undistorted shape of a galaxy. Arrows depict the direction of the gravitational force. If the field is tidal, as depicted in both cases, it exerts an opposite force on either side of the galaxy. On the left, the tidal field stretches the galaxy, while on the right it compresses it. This produces *intrinsic alignment*, where nearby galaxies in the same tidal field are oriented in the same direction.

The force on the matter that comprises the galaxy is the derivative of this potential, the gravitational force, depicted as arrows in Fig. 8.13. The first term, the constant potential, contributes no force on the galaxy; the second, the linear term, acts to move the galaxy uniformly in a single direction along the x-axis. It is the third term, the tidal force, the one that acts differently on the two sides of the galaxy, that distorts the shape. If $\phi''|_{x=0} > 0$, then the force $-d\phi/dx$ will be in the $-x$-direction on the right-hand side of the galaxy and in the other direction on the other side. The net result is the galaxy will be compressed along the x-direction, as shown in the right panel of the figure. If, on the other hand, ϕ'' is negative, then the tidal force will push both sides of the galaxy outwards, leading it to be elongated along the x-axis compared to its undistorted shape in the y-direction (assuming it were initially circular). In the language we have developed to describe shapes, the first case (when $\partial^2\phi/\partial x^2 > 0$) leads to $\epsilon_1 < 0$, while the second case leads to $\epsilon_1 > 0$. Therefore, we expect that

$$\epsilon_1 = -C\frac{\partial^2\phi}{\partial x^2} \tag{8.54}$$

with the coefficient $C > 0$. Moving beyond the one-dimensional variation in the potential, we can apply the same argument to the variation with respect to y, and combine to argue that we expect

$$\epsilon_1 = -C\left(\frac{\partial^2\phi}{\partial x^2} - \frac{\partial^2\phi}{\partial y^2}\right). \tag{8.55}$$

Not surprisingly, a 45° rotation of the image in Fig. 8.13 leads to a similar expression for ϵ_2:

$$\epsilon_2 = -2C\frac{\partial^2 \phi}{\partial x \partial y}. \tag{8.56}$$

Eqs. (8.55) and (8.56) appear very similar to our expressions for the ellipticity produced by gravitational lensing, so it is important to point out the deep differences between these equations and Eq. (8.2). The most important difference is that these equations purport to explain how the *matter* that comprises galaxies responds to local gravitational fields, while Eq. (8.2) describes how the *light* emitted from a background galaxy is distorted by structures along the line of sight. This difference feeds into the subtle, but crucial, difference between the potential in these two sets of equations. The potential here, ϕ in Eqs. (8.55) and (8.56), is the *local* 3D gravitational potential at the same radial distance from us as the galaxy of interest, while the potential in Eq. (8.2), Φ, is the projected 2D potential integrated along the full line of sight. Then, there is the important point that will have observational consequences: the signs are different. Intrinsic alignment leads a galaxy to be stretched where a lensed galaxy would be compressed (see Fig. 8.14). Finally, gravitational lensing, and therefore Eq. (8.2), follows directly from general relativity with no further assumptions. Our expressions for intrinsic alignment capture at best qualitatively (note the arbitrary constant in front) the impact of gravity on the shapes of galaxies, but they represent only a model. There are likely many other factors that

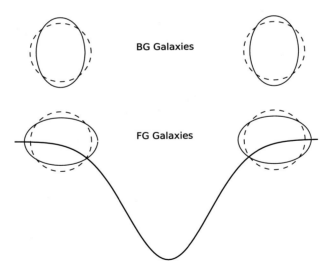

Figure 8.14 Two foreground (FG) galaxies situated on the outskirts of a potential well (depicted by the thick curve). The tidal forces stretch the galaxies so that if the horizontal direction in the figure were the x-axis, both galaxies would have $\epsilon_1 > 0$. This effect of intrinsic alignment is to be contrasted with gravitational lensing, wherein background galaxies (BG) at these same angular positions would appear tangentially sheared with $\epsilon_1 < 0$.

determine the shapes of galaxies: the history of their formation, electromagnetic processes, and nuclear energy are some of the obvious things missing.

The simplest manifestation of intrinsic alignment on the spectrum is the positive correlation of the two foreground galaxies in Fig. 8.14. The tidal forces acting on each will lead to $\langle \epsilon_1 \epsilon_1 \rangle > 0$ even in the absence of any gravitational lensing induced by structure along the line of sight. This will contaminate the signal of interest and lead to incorrect conclusions unless accounted for. To differentiate this contribution to the spectrum from the signal of gravitational lensing, we give each a name, with the intrinsic alignment part going by II and the gravitational lensing part by GG (i.e., $C_l^{GG} \equiv C_l^{\kappa\kappa}$). Note from Fig. 8.14 that the background galaxies are anti-aligned with the foreground galaxies: $\langle \epsilon_1^{BG} \epsilon_1^{FG} \rangle < 0$. This cross-term is denoted GI. The full spectrum then will be comprised of all three

$$C_l = C_l^{GG} + C_l^{GI} + C_l^{II}. \tag{8.57}$$

The intrinsic alignment spectra in Eq. (8.57) are usually subdominant to the GG lensing spectrum, but if not accounted for, they will lead to incorrect conclusions about the underlying cosmology. Therefore, it is important to make predictions for them and/or try to separate them out from the data. Employing the same steps that led to Eq. (8.40), we expect, for example,

$$C_l^{II} = \int dD_L \frac{[W^I(D_L)]^2}{D_L^2} P(l/D_L) \tag{8.58}$$

where $P(k)$ is the 3D matter power spectrum today and the window function is defined as

$$W^I(D_L) \equiv -\frac{3C\Omega_m H_0^2}{2c^2 D(z_L)} \frac{dn}{dD_L}. \tag{8.59}$$

The one new factor here is the *growth function* D, which depends on redshift (unfortunately for us, with all of the distances to keep track of, the symbol D is almost uniformly used for the growth function). You can work through an example in Exercise (8.3), but for now there are two important take-aways from the appearance of the growth factor in this equation. First, it is normalized to be equal to the scale factor a at early times, and indeed at high redshifts before dark energy began to dominate the universe, $D = a$. At late times, growth slows, and potential wells, which scale as D/a begin to decay. The interesting aspect of Eq. (8.59) is that we are dividing through by the growth function: this is an implicit implementation of the idea (which you can work through explicitly in Exercise (8.12)) that the orientations of galaxies are set by the tidal forces very early on, so the relevant potential is ϕ/D, the primordial potential.

Fig. 8.15 shows the predicted GG and II spectra for two different redshift distributions. Galaxies at very low redshifts are hardly sheared at all by intervening

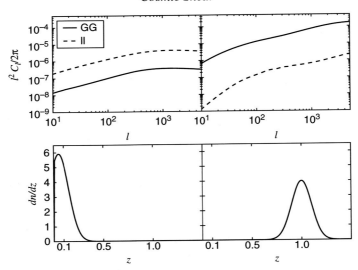

Figure 8.15 Predicted spectra for GG and II for two different redshift distributions. The left panels show the predicted signals from very low redshift galaxies; there is little cosmic shear since the light passes through only a very small part of the universe, so the intrinsic alignment signal dominates. For high redshift galaxies (right panels), the cosmic shear signal dominates, but intrinsic alignments still need to be accounted for to extract precise cosmological results.

matter because there is so little of it. The GG signal therefore is very small. The II signal, on the other hand, is much less sensitive to redshift, so at low redshifts it dominates. In fact, this is how intrinsic alignment has been detected: by looking for shape correlations in low redshift samples. This has helped constrain the amplitude C. The right panel in Fig. 8.15 shows that at high redshift the lensing signal dominates, but intrinsic alignment contaminates at the few percent level. As we move to the era of precision cosmology, it becomes increasingly important to account for this small contamination.

8.7 Tomography and Cosmological Parameters

By observing shapes of galaxies in multiple redshift bins, we can infer information about both the clustering of matter as a function of time and cosmological distances. This method is known as *tomography*, with the cartoon in Fig. 1.9 depicting the basic idea: the shapes of galaxies in the slice at z_2 are sensitive to structure at both z and z', while those in the slice at z_1 are distorted only by mass in the region z. We can now express this idea quantitatively by defining spectra that involve different background galaxy samples. Generalizing the κs in Eq. (8.32) to κ^i, the convergence inferred from a background galaxy sample with a distribution dn^i/dz, leads to

$$\langle \tilde{\kappa}^i(\vec{l})\kappa^j(\vec{l}')\rangle \equiv (2\pi)^2\delta_D^2(\vec{l}+\vec{l}')\,C_l^{\kappa\kappa,ij}. \tag{8.60}$$

For two redshift bins, there will be three spectra: 11, 22, and 12. More generally, if there are galaxies in N_z redshift bins, there will be $N_z(N_z + 1)/2$ spectra whose weights are determined by the two different redshift distributions:

$$C_l^{\kappa\kappa,ij} = \int_0^\infty dD_L \frac{W^i(D_L)W^j(D_L)}{D_L^2} P(l/D_L; a(D_L)) \qquad (8.61)$$

with

$$W^i(D_L) = \frac{2\Omega_m H_0^2 D_L}{2c^2 a(D_L)} \int_0^\infty dz \frac{dn^i}{dz} \frac{D_{SL}(z)}{D_S(z)} \Theta(D_S(z) - D_L). \qquad (8.62)$$

So, using the example in the cartoon in Fig. 1.9, if there are two background redshift bins at z_1 and z_2, there will be three measurable spectra: $C_l^{\kappa\kappa,11}$ will contain information about nearby structure; $C_l^{\kappa\kappa,22}$ about all structure near and far; and the cross-spectrum $C_l^{\kappa\kappa,12}$ will also probe only nearby structure but with a different weighting factor than $C_l^{\kappa\kappa,11}$.

Fig. 8.16 shows one example of how weak lensing tomography constrains cosmological parameters. The two weak lensing results shown are from two different surveys, the Dark Energy Survey (DES) and the Canada–France–Hawaii Telescope lensing project, that measured roughly the same size area and number of galaxies. In the case of DES, the constraints come from the spectrum measured shown in Fig. 8.12, albeit broken into three tomographic bins. The two parameters constrained are familiar to us with one caveat: Ω_m is the mean matter density in units

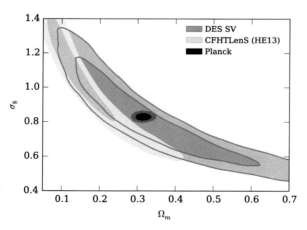

Figure 8.16 Constraints on cosmological parameters from two tomographic lensing surveys DES and CFHT (Abbott et al., 2016). The "SV" label on the DES results indicate that they come from Science Verification data, only 3 percent of the expected total data set; "HE13" refers to the paper that analyzed the CFHT data (Heymans et al., 2013). As of 2015, lensing constraints were still not as powerful as those from the cosmic microwave background, as indicated by the "Planck" allowed region. (See color plates section.)

of the critical density and σ_8 is the RMS of the density fluctuations, albeit with a twist. The subscript $_8$ refers to a filter of $8h^{-1}$ Mpc that is placed on the density field before its RMS is taken. You can show in Exercise (8.13) that this changes Eq. (8.22) to

$$\sigma_R^2 = \int_0^\infty \frac{dk}{k} \frac{k^3 P(k)}{2\pi^2} \left| \tilde{W}_R(k) \right|^2, \tag{8.63}$$

where R in this case is equal to $8h^{-1}$ Mpc and $\tilde{W}_R$ is the Fourier transform of the filter. Thus, σ_8 quantifies the RMS fluctuations on a particular scale. The lensing data that lead to the constraints in Fig. 8.16 can be explained either with a large value of Ω_m or a large fluctuation amplitude σ_8. This makes sense: the more matter there is in the universe, the larger will be the lensing signal. If the matter density is small, the same spectrum can still be produced if the fluctuations are large. The data depicted here are not sensitive enough to constrain the evolution of the fluctuations, but the hope is that upcoming data sets will be powerful enough to do so.

The cosmic shear tomographic measurements also have the potential to constrain cosmological distances. To see how, consider the situation where two redshift bins have dn^i/dz sharply peaked at $z_1 \simeq 0.3$ and dn^2/dz sharply peaked at $z_2 \simeq 0.7$. Then the integrands that determine W^i will be sharply peaked around z^i. Therefore, $D_S(z)$ in the integrand of Eq. (8.62) can be set to $D_S(z) \rightarrow D_S(z^i)$ and then the integral is equal to one, so that

$$W^i(D_L) \rightarrow \frac{2\Omega_m H_0^2}{2c^2 a(D_L)} \frac{D_{SL}^i D_L}{D_S^i} \Theta(D_S^i - D_L). \tag{8.64}$$

The step function then constrains the 11 spectrum to be sensitive to the power spectrum only at low z, while the W^{22} enables the spectrum to pick up contributions all the way out to z_2. The cross spectrum, $C_l^{\kappa\kappa,12}$ contains $\Theta(D_S^1 - D_L)$ so, like $C_l^{\kappa\kappa,11}$, is sensitive only to the power spectrum at low z; however, the weighting functions differ with

$$\frac{C_l^{\kappa\kappa,12}}{C_l^{\kappa\kappa,11}} \simeq \frac{\int_0^{D_S^1} \frac{dD_L}{a^2(D_L)} r^{12} P(l/D_L; a(D_L))}{\int_0^{D_S^1} \frac{dD_L}{a^2(D_L)} r^{11} P(l/D_L; a(D_L))} \tag{8.65}$$

with the distance ratio

$$r^{ij}(z_L) \equiv \frac{D_{SL}^i D_{SL}^j}{D_S^i D_S^j}. \tag{8.66}$$

Fig. 8.17 shows this distance weighting, and it uncovers something we should have anticipated: the distant galaxies are more affected by the same structure than are

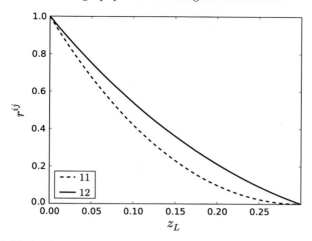

Figure 8.17 The ratio of distances as given in Eq. (8.66) that weight the integrand that determines the convergence auto-spectrum $C_l^{\kappa\kappa,11}$ and cross-spectrum $C_l^{\kappa\kappa,22}$ for source redshift bins sharply peaked at $z_1 = 0.3$ and $z_2 = 0.7$ as a function of lens redshift D_L. The cross-spectrum has a larger weighting because the background galaxies are more affected by foreground structure.

the nearby galaxies. We have seen that lenses close to sources have little impact on light propagation, so the $z_1 = 0.3$ galaxies do not get distorted by structure at $z = 0.29$ while the background $z_2 = 0.7$ galaxies feel the full impact. As a result, the cross-spectrum will be larger than the auto-spectrum and that difference will inform us about distance ratios.

Suggested Reading

To reinforce and go beyond the treatment of cosmology in the first few sections here, you might consult Ryden (2003); Coles and Lucchin (1995); Dodelson (2003b); Peacock (1999); or Weinberg (2008). Another important step to take is to get comfortable computing cosmological distances and spectra; a variety of software is available for this. I recommend the packages at `http://cosmologist.info/cosmomc/` and `https://bitbucket.org/joezuntz/cosmosis`.

There are several good reviews of dark energy: Frieman et al. (2008) and Caldwell and Kamionkowski (2009), with the former geared more to astronomers and the latter to physicists. As mentioned earlier, a nice review that focuses on what we can learn about dark energy from lensing is Hoekstra and Jain (2008).

The map in Fig. 8.6 and some of the subsequent constraints come from the Dark Energy Survey (Abbott et al., 2005), which took preliminary data in 2012–13 and began its official five years of operations in Fall 2013. It was designed to map 5000 square degrees so that the final mass map will be about 30 times larger than that shown in Fig. 8.6. Weinberg et al. (2013) provide an excellent, brief overview of all cosmic surveys planned for the years 2010–2030.

The derivation of the spectrum in §8.4 employs the *Limber Approximation* (Limber, 1953), used to compute the gravitational spectrum by Kaiser (1992). The approximation works well on small scales ($l \gtrsim 50$); as surveys get larger, there will be the need to go beyond Limber. A nice review on how to do this is in LoVerde and Afshordi (2008).

Figs. 8.13 and 8.14 and much of the subsequent discussion of intrinsic alignments is gleaned from the beautiful article by Catelan et al. (2001). More detailed models were introduced by Hirata and Seljak (2004) and reviewed by Troxel and Ishak (2015). Hu (1999) applied the word *tomography* to describe the idea of dividing background galaxies into different redshift bins and using their shapes to infer the evolution of structure and distances at different times.

Exercises

8.1 Using the values $\Omega_m = 0.315$ and assuming that the dark energy is a cosmological constant with $\Omega_{de} = 0.685$, construct the plot in Fig. 8.1. Also plot $\dot{a}/H_0 = aH/H_0$ and show that it is now increasing. This corresponds to the fact that the universe is currently accelerating.

8.2 Redo Exercise (8.1) for increasing values of dark energy equations of state w (all other parameters the same). Start with $w = -1$ corresponding to the cosmological constant, and then increase, using Eq. (8.10) and assuming that w is independent of z, until you find the value at which $\dot{a}$ never increases. This defines the critical value of w necessary to drive acceleration.

8.3 In a ΛCDM model, the growth function can be computed by carrying out the integral

$$D(a) = \frac{5\Omega_m H(a)}{2H_0} \int_0^a \frac{da'}{(a'H(a')/H_0)^3}. \tag{8.67}$$

Plot this growth function for the values of the parameters given in Exercise (8.1).

8.4 Write down Poisson's equation, (8.6), in Fourier space. Then relate the power spectrum of the 3D gravitational potential, P_ϕ, to the power spectrum of matter. Consider the values of the matter power spectrum $P(k)$ tabulated in Table 8.1, which are also shown in Figs. 8.4 and 8.5. Use them to plot $k^3 P_\phi/2\pi^2$ as a function of scale, as this quantifies the contributions to the variance of the gravitational potential.

Table 8.1 *Tabulated values of the power spectrum from the BOSS survey.*

$k\,(\mathrm{h\,Mpc^{-1}})$	$P(k)(\mathrm{Mpc}/h)^3$
0.00422	77867
0.0203	50617
0.0364	33715
0.0525	24685
0.0686	21561
0.0846	16308
0.101	12693
0.117	10858
0.1323	9191
0.149	7576
0.165	6240
0.181	5529
0.205	4387

8.5 Calculate the variance of the κ field in terms of $C_l^{\kappa\kappa}$ and show that it is equal to

$$\langle \kappa^2 \rangle = \int_0^\infty \frac{dl}{l} \frac{l^2 C_l^{\kappa\kappa}}{2\pi}. \tag{8.68}$$

8.6 The convergence is a weighted integral along the line of sight of the over-density field δ. As such, at a given radial position, we must evaluate δ not only at position $\vec{x}$, but also at time t corresponding to the time when the light from the source is passing by this position. Determine the redshift at this time as a function of distance from us. It is actually simpler to tabulate things in reverse: compute $D_L(z)$ as $\chi(z)/(1+z)$ where $\chi(z)$ is the integral in Eq. (7.26). Plot then $z(D_L)$.

8.7 The integral in Eq. (8.40) can also be written as an integral over wavenumber k, substituting dummy variables $k \equiv l/D_L$. Obtain an expression for $C_l^{\kappa\kappa}$ in terms of an integral over k.

8.8 Determine the redshift at which the power spectrum is most constrained by the cosmic shear spectrum for a given distribution of background galaxies. Take dn/dz to be a Gaussian function with mean $z = 1$ and standard deviation $\sigma_z = 0.2$ (not that different than the distribution shown in Fig. 8.9). Determine the weighting of the power spectrum in the integrand of Eq. (8.40): $W^2(D_L, D_S)/D_L$ (this is for the logarithmic integral $d\ln D_L$). Plotting this function as a function of redshift gives an indication of the redshifts to which the shear power spectrum is sensitive.

8.9 Consider a Gaussian distribution with mean zero and width σ:

$$G(x) = \frac{e^{-x^2/2\sigma^2}}{\sqrt{2\pi\sigma^2}}. \tag{8.69}$$

The mean of this 1D distribution is indeed zero, and the variance $\langle x^2 \rangle \equiv \int dx\, G(x) x^2$ is equal to σ^2. Show that the ensuing variance on the variance: $\langle (x^2 - \sigma^2)^2 \rangle = 2\sigma^4$.

8.10 Now imagine computing the variance of the Gaussian distribution described in Exercise (8.9) using N data points x_i with $i = 1, \ldots, N$. Since each data point is drawn from the same distribution, its expected mean is zero and $\langle x_i^2 \rangle = \sigma^2$. Determine the variance of the estimator for the variance:

$$\hat{\sigma}^2 \equiv \frac{1}{N} \sum_i x_i^2. \tag{8.70}$$

by showing that

$$\mathrm{Var}\left(\hat{\sigma}^2\right) \equiv \left\langle \left(\hat{\sigma}^2 - \sigma^2\right)^2 \right\rangle = \frac{2\sigma^4}{N}. \tag{8.71}$$

8.11 Estimate $\Delta C_l^{\kappa\kappa}$ using Eq. (8.50) with $f_{\mathrm{sky}} = 0.0034$ and galaxy density of four per square arcminute. Show that they are roughly equal to the heights of the rectangles in Fig. 8.12. Note that each measurement averages over a bandwidth of $\Delta l \sim 150$ so you will need to divide the expression in Eq. (8.50) by the square of this bandwidth.

8.12 Derive Eq. (8.58). Make use of Poisson's equation and also assume that the ellipticities given in Eqs. (8.55) and (8.56) are set by the primordial 3D potential very early on. This means that, to get to Eq. (8.58), which is in terms of the matter power spectrum today, you must extrapolate from the primordial power spectrum to the present-day value. Hint: that explains the factors of the growth function in the denominator.

8.13 Consider the overdensity field filtered on a scale R such that

$$\delta_R(\vec{x}) \equiv \int d^3x'\, W_R\left(|\vec{x} - \vec{x}'|\right) \delta(\vec{x}'). \tag{8.72}$$

Show that the RMS fluctuations in this filtered field are given by Eq. (8.63). If the filtering function W is a top-hat, so that $W_R = 1$ if its argument is less than R but equal to zero otherwise, determine the Fourier transform $\tilde{W}_R(k)$.

8.14 The distance ratio defined in Eq. (8.66) for two tomographic bins is plotted in Fig. 8.17 as a function of the lens redshift when the two source bins are at $D_S = 0.3$ and $D_S = 0.7$. Reproduce the solid curve in this plot using $\Omega_m = 0.31$ and $h = 0.68$. Then, show how the ratio changes when $\Omega_m \rightarrow 0.2$ and 0.5 to get a sense of how sensitive lensing is to cosmological distances and therefore the parameters that determine them. In all cases, assume that the dark energy is a cosmological constant with value such that $\Omega_m + \Omega_\Lambda = 1$.

9

Lensing of the Cosmic Microwave Background

The Cosmic Microwave Background (CMB) is a relic of the universe when it was only 380,000 years old. Standard lore has it that the photons in the CMB have traveled freely to us since that time, so constitute an unvarnished picture – literally a picture – of the early universe. We have already seen some of the power of this picture, in that the constraints on cosmological parameters from the CMB, as shown for example in Fig. 8.16, are extraordinarily tight.

The purpose of this chapter is to remove one word from the preceding paragraph: "unvarnished." In fact, the photons in the CMB do **not** travel completely freely to us; rather, their paths are distorted by the gravitational potentials along the line of sight. On the simplest level this means that CMB photons observed as emanating from a given direction actually started their 13-billion-year journey from a slightly different direction. We have seen this through the book and learned to ask: how can the deflections be observed?

Understanding the effect of CMB lensing, as these deflections are called, is not as simple as understanding how a point source is magnified or a galaxy is stretched. Rather, it is the statistics of the CMB that are distorted. So we have to rely on our growing understanding of the spectrum of fields like the CMB and observe the small distortions in the spectrum in order to infer information about the projected gravitational potential.

9.1 The Cosmic Microwave Background

Photons characterized by a blackbody distribution with temperature $T = 2.725\text{K}$ permeate the universe. The peak of the intensity of a blackbody boson gas is at $\nu = 3k_B T/h$, where k_B is the Boltzmann constant and h Planck's constant. In this case the peak is at 160 GHz, in the microwave region of the electromagnetic spectrum. So there is a *microwave background*. This radiation is a relic of the early universe and conveys information about the state of the universe at early times. A photon with a frequency today of 160 GHz would have had a larger frequency (shorter wavelength) earlier in the history of the universe, when the scale factor that dictates physical lengths was smaller. Going back to the epoch when the typical frequency of the background photons was in the optical regime with energies large enough to ionize hydrogen, we can imagine that the early universe was likely ionized with only free protons and electrons and essentially no neutral hydrogen. The ionization was extraordinarily effective because the number of photons is so much larger than the number of electrons and protons.[1] The three components were tightly coupled with one another and formed a fluid with a common temperature, because free electrons scatter often off of photons via Compton scattering and off of protons via Coulomb scattering. The tight coupling meant that any over-dense region contained a surplus of all components of the fluid: a region with a temperature slightly larger than average would contain more photons and more electrons and more protons than an average region.

As the universe expanded and the scale factor increased, the temperature of this fluid cooled until eventually the average photon energy was too small to ionize hydrogen. Since there were/are many more photons than electrons (Exercise (9.4)), there were still enough high-energy photons in the tail of the distribution to ionize hydrogen even at relatively low temperatures. Eventually, though, when the temperature dropped to 3000K (corresponding to an energy a tenth that of the ionization energy of hydrogen), electrons and protons combined to form neutral hydrogen, a process dubbed *recombination*. After that, all photons in the background ceased to interact with matter and traveled freely through the universe. So, when we observe them today, they are clearly part of a *cosmic microwave background* (CMB) and they convey information about the state of the universe when the temperature was about 3000K, corresponding to a redshift $z \simeq 1100$. The CMB then provides a picture of the universe when it was very young.

Since photons released at that time close to us have already passed us and photons released very far away have not yet reached us, we observe photons that originated from a spherical shell. This shell is called the surface of last scattering.

The overarching sense we get of the universe from this picture (Fig. 1.10) is that the electrons and photons and protons were distributed very homogeneously. In contrast to today, when the density in our Galaxy is five orders of magnitude larger than in a random place in the Universe, the density in the early universe was homogeneous to one part in ten thousand. If one region had ten thousand protons, another might have 9,999, but no region has 9,800 or even 9,950. This follows directly from the amplitude of the hot and cold spots in Fig. 1.10: since the density scales as T^4,

---

[1] The number of electrons is equal to the number of protons as the total electric charge of the universe is zero.

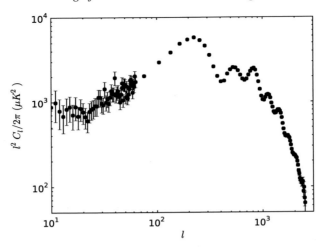

Figure 9.1 Spectrum of anisotropies in the cosmic microwave background as measured by the Planck satellite. Data Credit: ESA

a fractional excess temperature of $80\,\mu K/2.7K = 3 \times 10^{-5}$, which corresponds to a density contract of 1.2×10^{-4}, a part in ten thousand.

Using the same techniques of Fourier transforms and power spectra that described weak lensing, we can analyze these anisotropies by plotting their spectrum. This is shown in Fig. 9.1 with the features observed in the map apparent from the vertical and horizontal scales in the figure. The hot and cold spots have typical sizes of order a degree, and this corresponds to the first peak at multipole $l \sim \pi/1° \sim 200$ as seen in the figure. The RMS amplitude of these spots is $80\mu K$, so the variance at the peak is roughly $6400\mu K^2$, as indicated by the spectrum.

The shape of the anisotropy spectrum conveys information about the underlying physics. When the photons, electrons, and protons were tightly coupled, they moved together. The fluid would migrate out of an initially overdense region due to its pressure, dominated by the relativistic photons. The result was that, at a given place, the density oscillated between lower than and greater than the mean density. The CMB is a snapshot of those spatially varying oscillations at the time of recombination. These oscillations are due to the restoring force of pressure, just as the string on a guitar is subject to a restoring force. Hence, the fluctuations in the early universe fluid and the height of a guitar string are both governed by the wave equation. Just as the spectrum of a musical instrument displays a fundamental mode and higher harmonics, the spectrum of the CMB has its first peak at $l \sim 200$ and then subsequent peaks at integer multiples of the fundamental mode. The other feature common with musical instruments is the damping of the spectrum at higher harmonics, corresponding to small spatial scales. In this case, the physics that drives the damping is that photons can random walk out of over-dense regions

by scattering off of electrons (this effect goes beyond the single fluid approximation). This effect is quite apparent in Fig. 9.1, as the amplitude of the fluctuations
drops by a factor of ten going from $l = 200 \rightarrow l = 2000$. It will turn out to have
important implications for the lensing of the CMB.

Finally, at recombination, the correlations were homogeneous, an attribute that
we saw on page 170 leads to a constraint on the spectrum of the anisotropies:

$$\langle \tilde{T}(\vec{l})\tilde{T}(\vec{l}')\rangle = (2\pi)^2 \, \delta_D^2(\vec{l} + \vec{l}') \, C_l. \tag{9.1}$$

That is, the primordial fluctuations of modes with different $\vec{l}$ were uncorrelated with
one another. Two minor comments about Eq. (9.1), which will form the basis for
this chapter. First, although we have seen several instances already of 2D spectra
labeled C_l, we will reserve the un-superscripted C_l for the (unlensed) CMB spectrum. Second, it will prove useful to write the product on the left with the complex
conjugate of one of the factors of $\tilde{T}$ so that Eq. (9.1) becomes

$$\langle \tilde{T}(\vec{l})\tilde{T}^*(\vec{l}')\rangle = (2\pi)^2 \, \delta_D^2(\vec{l} - \vec{l}') \, C_l \,. \tag{9.2}$$

Here we have used the fact that you can show in Exercise (9.5): the Fourier trans-
form of a real function satisfies $\tilde{T}(\vec{l}') = \tilde{T}^*(-\vec{l}')$, so that the sign in the argument
of the Dirac delta function changes.

9.2 Order of Magnitude Estimate

We are interested in understanding the impact of the deflections experienced by
the photons in the CMB as they travel to us from the surface of last scattering. For
orientation, we will find that the photons in the CMB have been deflected by a few
arcminutes, but the structures that caused the deflections were quite large, of order
a degree or larger. This leads to the counterintuitive result that in order to measure
the large-scale structure responsible for deflection, one needs to measure the CMB
on small angular scales.

To see this, consider the variance of the deflection angle $\vec{\alpha} = \vec{\nabla}\Phi$. In Fourier
space, we know we can write this as

$$\begin{aligned}
\alpha_{rms}^2 &= \int \frac{d^2l}{(2\pi)^2} \, C_l^{\nabla\Phi} \\
&= \int_0^\infty \frac{dl}{l} \, \frac{l^4 C_l^\Phi}{2\pi}
\end{aligned} \tag{9.3}$$

where the second line follows since $\nabla \rightarrow il$ in Fourier space. The combination
$l^4 C_l^\Phi/2\pi$ is related to $C_l^{\kappa\kappa}$, which was introduced in Chapter 8, with the latter containing an extra two powers of l and down by a factor of $1/4$. That is, the variance of
the deflection angle gets contributions from angular scale l of $(4/l^2)\,(l^2 C_l^{\kappa\kappa}/2\pi)$.

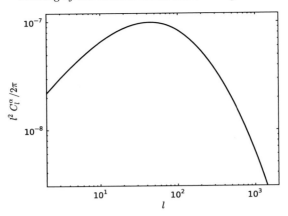

Figure 9.2 Angular power spectrum of the deflection angle α of CMB photons. The deflections are dominated by $l \sim 50$ scale structures ($\sim$ a hundred Mpc) and have an RMS amplitude of order ($10^{-7/2} \sim 1'$).

This means that it peaks at a value $l_{max} \simeq 50$, as shown in Fig. 9.2, and that the amplitude is a factor of $4/2500$ smaller than the variance of κ. The first of these observations does indeed translate into the statement that most of the deflection is due to structures at $l \sim 50$, a few degrees on the sky. Taking the typical cosmological distance out to the last scattering surface as a few Gpc, this corresponds to structures of order 100 Mpc. The root mean square fluctuations (the square root of the variance) in κ back to the surface of last scattering are a little less than 0.01, so the RMS fluctuations in the deflection angle are twenty-five times smaller, of order $3 \times 10^{-4} \simeq 1'$.

9.3 Impact of Convergence on the Acoustic Peaks

When the CMB photons pass by a region of nonzero convergence, their paths are altered in ways that we have come to understand, starting from Fig. 4.2. If we approximate the hot and cold spots in Fig. 1.10 as circles on the sky, then photons from a hot spot at the last scattering surface passing through a region of $\kappa > 0$ transform the circle so that it appears larger. If its initial angular radius was θ, its observed radius will be $\theta' = \theta(1 + \kappa)$. Physically, then, we expect features in the spectrum such as peaks and troughs to move to larger scales when the line of sight κ is positive. Since multipole l is conjugate to angular scale, this means the spectrum depicted in Fig. 9.1 should shift to the left when $\kappa > 0$.

More quantitatively, the spectrum is related to the variance of the Fourier transform of the temperature, $\tilde{T}(\vec{l})$. If we denote the lensed temperature simply as T and the unlensed temperature as T^u, then in this toy example of a circular hot spot, $T(\vec{\theta}) = T^u(\vec{\theta}/(1 + \kappa))$. The observed Fourier transform therefore is

$$\tilde{T}(\vec{l}) = \int d^2\theta \, e^{-i\vec{l}\cdot\vec{\theta}} \, T^{\mathrm{u}}(\vec{\theta}/(1+\kappa)). \tag{9.4}$$

Now change dummy variables to $\vec{\theta}' \equiv \vec{\theta}/(1+\kappa)$, so that

$$\tilde{T}(\vec{l}) = (1+\kappa)^2 \int d^2\theta' \, e^{-i\vec{l}\cdot\vec{\theta}'(1+\kappa)} \, T^{\mathrm{u}}(\vec{\theta}') \tag{9.5}$$

where the prefactor comes from the Jacobian transforming from one dummy variable θ to the other θ'. We can absorb the factor of $1 + \kappa$ in the exponential into $l' \equiv l(1+\kappa)$, so that

$$\tilde{T}(\vec{l}) = (1+\kappa)^2 \, \tilde{T}^{\mathrm{u}}\left(\vec{l}(1+\kappa)\right). \tag{9.6}$$

With this expression for the observed Fourier transform in terms of the unlensed temperature, whose spectrum is known, we can calculate the observed spectrum behind a region with nonzero κ in terms of the unlensed spectrum. As usual, we take the average of the product of two Fourier transforms:

$$\langle \tilde{T}(\vec{l}) \, \tilde{T}^*(\vec{l}')\rangle = (1+\kappa)^4 \, (2\pi)^2 \, \delta_D^2\left(\vec{l}[1+\kappa] - \vec{l}'[1+\kappa]\right) C_{l(1+\kappa)} \tag{9.7}$$

where the C_l on the right-hand side is the unlensed spectrum. Remembering that transforming the Dirac delta function into the standard $\delta_D^2(\vec{l} - \vec{l}')$ brings down a factor of $(1+\kappa)^{-2}$ leads to a simple expression for the spectrum behind a region with nonzero kappa:

$$
\begin{aligned}
C_l^{\mathrm{obs}}(\kappa) &= (1+\kappa)^2 \, C_{l(1+\kappa)} \\
&\simeq C_l \left[1 + 2\kappa + \frac{1}{C_l}\frac{dC_l}{dl} \, l\kappa\right]
\end{aligned} \tag{9.8}
$$

where the second line follows from a first-order Taylor expansion in κ. We can collect the two terms linear in κ to write the fractional change in the spectrum due to κ as

$$\frac{\Delta C_l(\kappa)}{C_l} = \kappa\frac{d\ln(l^2 C_l)}{d\ln l}. \tag{9.9}$$

Fig. 9.3 shows the impact of this change for two different values of κ. Qualitatively, it is as we expected: for positive κ the spectrum shifts to the left. Quantitatively, we see that the change in the power for regions behind positive κ is positive if the power $l^2 C_l$ is rising and negative if it is falling. The error bar in the figure shows that projected from a one square degree region imaged by a high-resolution CMB experiment. Even averaging over ~ 10 such bins would not lead to a detection: the signal is small. However, as we will see in the next two sections, the effects of CMB lensing can be extracted from data using clever analysis techniques.

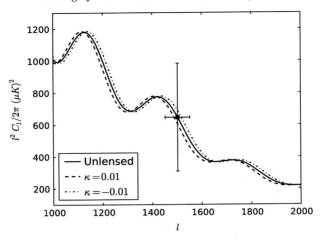

Figure 9.3 Blow-up of the region around the small-scale acoustic peaks in the CMB anisotropy spectrum. Dashed curve shows that when CMB photons traverse a region with $\kappa = +0.01$, the hot and cold spots are enlarged and hence the peak moves to lower ls. Dot-dashed curve shows that the spots shrink and therefore move to higher l when the photons traverse an underdense region. The error bar is projected from a single one square degree region; even including ten such bins from $1000 < l < 2000$ would decrease the error by only $\sqrt{10}$, so the signal induced by κ is difficult to detect.

9.4 Inferring the Lensing Potential

We have learned something interesting about the CMB: it will not have the same statistics in different regions of the sky. Experiments targeting a given direction will measure one anisotropy spectrum, while experiments targeting a different region with a different value of κ will measure a different spectrum. Equivalently, the correlation function $\langle T(\vec{\theta}_1)T(\vec{\theta}_2)\rangle$ will depend not only on the angular distance $|\vec{\theta}_1 - \vec{\theta}_2|$ but also on the location on the sky $(\vec{\theta}_1 + \vec{\theta}_2)/2$ say. It is useful to go back to §8.2 to remind ourselves of the implications of this dependence on position (the inhomogeneity of the statistics). When the correlation function does not depend on position, we found that Fourier modes with different wave numbers were uncorrelated, which would translate in our case into the statement $\langle \tilde{T}(\vec{l})\tilde{T}^*(\vec{l}')\rangle$ is proportional to a Dirac delta function with argument $\vec{l} - \vec{l}'$. The inhomogeneity induced by the convergence will lead to a breakdown in this relation, so that modes with different values of $\vec{l}$ will be correlated.

Let us demonstrate this by taking the product of the lensed temperatures at two different values of $\vec{l}$. If the field were homogeneous, the expectation value would vanish. But lensing means some lines of sight are different from others; that the universe is not homogeneous. This means that Fourier modes with different wave numbers **will** be correlated, and that we can hope to learn about the inhomogeneity – that is, about the lensing field κ – by studying these correlations. We first need to calculate the Fourier transform of the observed temperature field as distorted by κ, moving beyond the simple toy example of the last section where a circular hot spot was dilated or contracted. The starting point is the realization that the observed temperature at position $\vec{\theta}$ is equal to the unlensed temperature at position $\vec{\theta} - \vec{\alpha}$. We can Taylor expand this equality in the small deflection angle $\vec{\alpha}$:

$$T(\vec{\theta}) = T^{\mathrm{u}}(\vec{\theta} - \vec{\alpha})$$

$$= T^{\mathrm{u}}(\vec{\theta}) - \frac{\partial T^{\mathrm{u}}}{\partial \theta_i} \alpha_i + \frac{1}{2} \frac{\partial^2 T^{\mathrm{u}}}{\partial \theta_i \partial \theta_j} \alpha_i \alpha_j + \dots \tag{9.10}$$

For now, we need keep only the linear term, although in the next section the quadratic term will also prove relevant. Taking the Fourier transform of the first two terms leads to

$$\tilde{T}(\vec{l}) \simeq \tilde{T}^{\mathrm{u}}(\vec{l}) - \frac{1}{c^2} \int d^2\theta e^{-i\vec{l}\cdot\vec{\theta}} \int \frac{d^2 l_1}{(2\pi)^2} e^{i\vec{l}_1\cdot\vec{\theta}} i\vec{l}_1 \tilde{T}^{\mathrm{u}}(\vec{l}_1) \int \frac{d^2 l_2}{(2\pi)^2} e^{i\vec{l}_2\cdot\vec{\theta}} i\vec{l}_2 \tilde{\Phi}(\vec{l}_2) \tag{9.11}$$

where we have used the relation between potential and deflection angle: $\vec{\alpha} = (1/c^2)\nabla\Phi$. The $d^2\theta$ integral leads to a Dirac delta function; using it to collapse the l_2 integral in the linear term leads to

$$\tilde{T}(\vec{l}) \simeq \tilde{T}^{\mathrm{u}}(\vec{l}) + \frac{1}{c^2} \int \frac{d^2 l_1}{(2\pi)^2} \vec{l}_1 \cdot (\vec{l} - \vec{l}_1)\tilde{T}^{\mathrm{u}}(\vec{l}_1) \, \tilde{\Phi}(\vec{l} - \vec{l}_1). \tag{9.12}$$

Comparing the second terms in Eqs. (9.10) and (9.12), we see a very general feature of Fourier transforms. A product in real space ($\frac{\partial T^{\mathrm{u}}}{\partial \theta_i} \times \alpha_i$) becomes a convolution in Fourier space and vice versa, as you can show in Exercise (9.7).

We can now calculate the correlation between two different $\vec{l}$ modes induced by lensing. Consider

$$\langle \tilde{T}(\vec{l})\tilde{T}^*(\vec{l}')\rangle_{\vec{l}\neq\vec{l}'} \simeq \frac{1}{c^2} \int \frac{d^2 l_1}{(2\pi)^2} \vec{l}_1 \cdot (\vec{l}-\vec{l}_1) \tilde{\Phi}(\vec{l}-\vec{l}_1)\langle \tilde{T}^{\mathrm{u}}(\vec{l}')^*\tilde{T}^{\mathrm{u}}(\vec{l}_1)\rangle + c.c. \tag{9.13}$$

where c.c. stands for complex conjugate and the average denoted by $\langle \dots \rangle$ is over realizations of the unlensed temperature field for fixed projected gravitational potential Φ. This average then yields a Dirac delta function multiplied by $C_{l'}$, so

$$\langle \tilde{T}(\vec{l})\tilde{T}^*(\vec{l}')\rangle_{\vec{l}\neq\vec{l}'} \simeq \frac{1}{c^2} C_{l'} \vec{l}' \cdot (\vec{l} - \vec{l}')\tilde{\Phi}(\vec{l} - \vec{l}') + c.c. \tag{9.14}$$

The idea is that, for fixed $\tilde{\Phi}$, the product of the two temperatures will on average give back the correct $\tilde{\Phi}$, weighted by the various factors of $\vec{l}, \vec{l}'$. Defining $\vec{L} \equiv \vec{l} - \vec{l}'$ then leads to a simple *quadratic estimator* for the Fourier transform of the potential

$$\hat{\tilde{\Phi}}(\vec{L}) = c^2 \frac{\tilde{T}(\vec{l})\tilde{T}^*(\vec{l}-\vec{L})}{C_{\vec{l}-\vec{L}}\vec{L}\cdot(\vec{l}-\vec{L})}. \tag{9.15}$$

This estimator combines two Fourier small-scale modes, each with wavenumber pretty close to one another, separated only by the small vector $\vec{L}$.

Eq. (9.15) is not the best way to infer the potential. First, it uses only a single small-scale mode $\vec{l}$; obviously, it makes sense to combine information from many small-scale modes, in each case taking the product of two temperatures with arguments separated by $\vec{L}$. Second, the estimator in Eq. (9.15) will indeed retrieve the true value of $\tilde{\Phi}$ on average but will be very noisy. The quadratic terms can be weighted to reduce the noise while retaining the attractive feature that the expected value of the estimator is equal to the true potential.

Fig. 9.4 shows an all-sky map of the projected potential using the information contained in the temperature field. This is a remarkable map: unlike most astronomical pictures, it is not an inventory of the light received from a source or sources. Rather, it is, like Fig. 8.6, a map of the mass in the different directions on the sky. In some sense, it is the ultimate mass map, for it includes all mass back to the surface of last scattering. On the other hand, not all of the features evident in the map are signal: the signal-to-noise of each spot is of order one, so any one feature could be noise. Statistically, though, the total signal-to-noise ratio of the all-sky map is over 25-sigma.

There is room to do even better: ground-based experiments have made higher signal-to-noise maps on smaller areas of the sky. Apart from improving the fidelity of the maps, there is a large effort underway to use these maps, comparing to those taken in the standard way by tracing the light emitted from sources. How much information can be gleaned from cross-correlating these different maps with one another is an active area of research. Taken together, all of

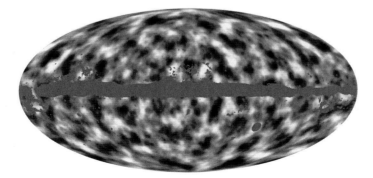

Figure 9.4 A map of the projected gravitational potential inferred from a quadratic estimator of the CMB temperature by the Planck satellite. Copyright: ESA and the Planck Collaboration. (See color plates section.)

these challenges make CMB lensing one of the most fertile areas of research in cosmology.

9.5 Smoothing the Peaks and Troughs

Another ramification of CMB lensing is the smoothing of the acoustic peaks and troughs in the anisotropy spectrum. Eq. (9.9) expresses the change in the spectrum in a small region of sky over which κ is constant. When averaged over the entire sky, this effect would appear to cancel out since there are as many negative as positive regions of κ. However, a more careful treatment that includes second-order terms will illustrate that the peaks and troughs in the spectrum are smoothed out.

Therefore, the effect of lensing on the CMB spectrum requires us to keep terms quadratic in κ. Returning to Eq. (9.10), we need to retain the third term so that the Fourier transform becomes

$$
\begin{aligned}
\tilde{T}(\vec{l}) = \tilde{T}^u(\vec{l}) &- \frac{1}{c^2} \int d^2\theta e^{-i\vec{l}\cdot\vec{\theta}} \left[\int \frac{d^2 l_1}{(2\pi)^2} e^{i\vec{l}_1\cdot\vec{\theta}} i\vec{l}_1 \tilde{T}^u(\vec{l}_1) \int \frac{d^2 l_2}{(2\pi)^2} e^{i\vec{l}_2\cdot\vec{\theta}} i\vec{l}_2 \tilde{\Phi}(\vec{l}_2) \right. \\
&+ \frac{1}{2c^4} \int \frac{d^2 l_1}{(2\pi)^2} e^{i\vec{l}_1\cdot\vec{\theta}} l_{1i}l_{1j}\tilde{T}^u(\vec{l}_1) \int \frac{d^2 l_2}{(2\pi)^2} e^{i\vec{l}_2\cdot\vec{\theta}} l_{2i}\tilde{\Phi}(\vec{l}_2) \\
&\left. \times \int \frac{d^2 l_3}{(2\pi)^2} e^{i\vec{l}_3\cdot\vec{\theta}} l_{3j}\tilde{\Phi}(\vec{l}_3) \right].
\end{aligned}
\tag{9.16}
$$

The θ integral leads to a Dirac delta function; using it to collapse the l_2 integral in the linear term and the l_3 integral in the quadratic term leads to

$$
\begin{aligned}
\tilde{T}(\vec{l}) = \tilde{T}^u(\vec{l}) &+ \frac{1}{c^2} \int \frac{d^2 l_1}{(2\pi)^2} \vec{l}_1 \cdot (\vec{l}-\vec{l}_1)\tilde{T}^u(\vec{l}_1)\,\tilde{\Phi}(\vec{l}-\vec{l}_1) \\
&+ \frac{1}{2c^4} \int \frac{d^2 l_1}{(2\pi)^2} \tilde{T}^u(\vec{l}_1) \int \frac{d^2 l_2}{(2\pi)^2} (\vec{l}_1\cdot\vec{l}_2)\vec{l}_1 \cdot (\vec{l}-\vec{l}_1-\vec{l}_2) \\
&\times \tilde{\Phi}(\vec{l}_2)\tilde{\Phi}(\vec{l}-\vec{l}_1-\vec{l}_2).
\end{aligned}
\tag{9.17}
$$

To obtain the effect of lensing on the power spectrum, we multiply two factors of $\tilde{T}$ together, take the expectation value over both temperature and potential, and keep only terms that are second-order in $\tilde{\Phi}$.

$$
\langle \tilde{T}(\vec{l})\,\tilde{T}(\vec{l})^* \rangle = (2\pi)^2\,\delta_D^2(\vec{l}-\vec{l}')C_l + FF + ZS
\tag{9.18}
$$

where the zeroth-order term is the unlensed spectrum; the product of the two terms of first order in Φ is defined as

$$FF \equiv \frac{1}{c^4} \int \frac{d^2l_1}{(2\pi)^2} \vec{l}_1 \cdot (\vec{l} - \vec{l}_1) \int \frac{d^2l_1'}{(2\pi)^2} \vec{l}_1' \cdot (\vec{l}' - \vec{l}_1')$$
$$\times \langle \tilde{T}^u(\vec{l}_1) \, \tilde{T}^u(\vec{l}_1')^* \rangle \langle \tilde{\Phi}(\vec{l} - \vec{l}_1) \tilde{\Phi}^*(\vec{l}' - \vec{l}_1') \rangle; \tag{9.19}$$

and the zeroth-order term multiplying the second-order term is defined as

$$ZS \equiv 2\frac{1}{2c^4} \int \frac{d^2l_1}{(2\pi)^2} \int \frac{d^2l_2}{(2\pi)^2} (\vec{l}_1 \cdot \vec{l}_2) \vec{l}_1 \cdot (\vec{l}' - \vec{l}_1 - \vec{l}_2)$$
$$\times \langle \tilde{T}^u(\vec{l}) \tilde{T}^u(\vec{l}_1)^* \rangle \langle \tilde{\Phi}^*(\vec{l}_2) \tilde{\Phi}^*(\vec{l}' - \vec{l}_1 - \vec{l}_2) \rangle. \tag{9.20}$$

The factor of two in front of this last term is due to the zeroth-order part of $\tilde{T}(\vec{l})$ multiplying the second-order piece of $\tilde{T}(\vec{l}')^*$ added to the second-order part of $\tilde{T}(\vec{l})$ multiplying the zeroth piece of $\tilde{T}(\vec{l}')^*$ (both give the same result).

First let's reduce the FF term by noting that each of the expectation values lead to Dirac delta functions; these force $\vec{l}_1 = \vec{l}_1'$, thereby eliminating the d^2l_1' integral and then $\vec{l} = \vec{l}'$. They also bring in the relevant power spectra: C_{l_1} and $C^\Phi_{\vec{l}-\vec{l}_1}$. Therefore

$$FF = \frac{(2\pi)^2 \delta_D^2(\vec{l} - \vec{l}')}{c^4} \int \frac{d^2l_1}{(2\pi)^2} C_{l_1} \left[\vec{l}_1 \cdot (\vec{l} - \vec{l}_1) \right]^2 C^\Phi_{\vec{l}-\vec{l}_1}. \tag{9.21}$$

We can now use some physics: we know that the lensing will be done by large-scale structures, so the argument of the power spectrum of the potential, $\vec{l} - \vec{l}_1$, will be quite small. Therefore, the integral over $\vec{l}_1$ will get most of its contribution when $\vec{l}_1$ is very close to $\vec{l}$. The CMB spectrum in the integrand C_{l_1} will be essentially equal to C_l; more precisely it will be equal to the spectrum smoothed over a region $l - \Delta l \rightarrow l + \Delta l$ with $\Delta l \simeq 50$ (since that is the scale at which the lensing spectrum peaks as shown in Fig. 9.2). Let's call that smoothed CMB spectrum $\bar{C}_l$. Then, extracting the piece proportional to the Dirac delta function, we can identify this first correction to the lensing spectrum:

$$\Delta C_l^{FF} = \bar{C}_l \int \frac{d^2l_1}{(2\pi)^2} \left[\vec{l}_1 \cdot (\vec{l} - \vec{l}_1) \right]^2 \frac{C^\Phi_{\vec{l}-\vec{l}_1}}{c^4}. \tag{9.22}$$

Finally, it pays to write this a bit more compactly by redefining the dummy variable to be $\vec{l}' \equiv \vec{l} - \vec{l}_1$ so that

$$\Delta C_l^{FF} = \bar{C}_l \int \frac{d^2l'}{(2\pi)^2} (\vec{l} \cdot \vec{l}')^2 \frac{C^\Phi_{l'}}{c^4} \tag{9.23}$$

where the fact that $\vec{l} \gg \vec{l}'$ has been used.

The cross term in Eq. (9.20) can be treated similarly; the second expectation value in Eq. (9.20) sets $\vec{l}_1 = \vec{l}'$, which is also equal to $\vec{l}$ by virtue of the first expectation value. Therefore,

$$ZS = -C_l \frac{(2\pi)^2 \delta_D^2(\vec{l} - \vec{l}')}{c^4} \int \frac{d^2 l_2}{(2\pi)^2} (\vec{l} \cdot \vec{l}_2)^2 C_{l_2}^\Phi. \tag{9.24}$$

We can extract the change in the spectrum due to this term, and it looks remarkably similar to Eq. (9.23):

$$\Delta C_l^{ZS} = -C_l \int \frac{d^2 l'}{(2\pi)^2} (\vec{l} \cdot \vec{l}')^2 \frac{C_{l'}^\Phi}{c^4}. \tag{9.25}$$

The only differences are the sign and that the prefactor in one case is the smoothed C_ls, while it is the raw spectrum in the other case.

We can now add the two contributions to find the change in the CMB spectrum due to lensing:

$$\Delta C_l = (\bar{C}_l - C_l) \int \frac{d^2 l'}{(2\pi)^2} (\vec{l} \cdot \vec{l}')^2 \frac{C_{l'}^\Phi}{c^4}. \tag{9.26}$$

The first point about this expression is that it predicts a smoothing of the peaks and troughs. At a peak, the correction factor is negative since C_l is greater than its smoothed average $\bar{C}_l$; at a trough, it is positive. Therefore, the effect of lensing on the C_ls is to smooth out the spectrum. This effect, depicted in Fig. 9.5, has been detected by several high-resolution CMB experiments.

The more quantitative point is that the amplitude of this smoothing does indeed depend on $l^2 C_l^\alpha$, the power in the deflection angle (Exercise (9.8)). This amplitude

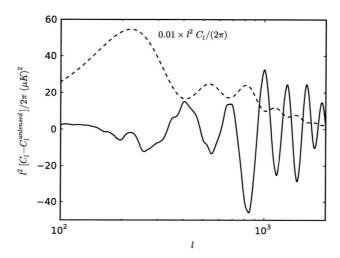

Figure 9.5 Difference between the lensed and unlensed CMB anisotropy spectrum as a function of angular scale. The absolute spectrum divided by 100 (so it appears on the same scale) is depicted by the dashed curve. Note that lensing smooths the peaks and troughs: the heights of the peaks are lower than the unlensed spectrum and the heights of the troughs are higher.

can be measured and compared with the prediction, as any cosmological model will lead to a prediction for C_l^Φ (or equivalently C_l^α).

Suggested Reading

The era of the CMB, and arguably modern cosmology, began in 1965 with the two papers by Dicke et al. (1965) and Penzias and Wilson (1965). Both articles are short enough to read in an hour, and are well worth the effort. A nice compilation of essays by some of the pioneers in the field of CMB research is collected in Peebles et al. (2009). The story of the peaks and troughs in the CMB is more nuanced than is explained in §9.1. As explained in Dodelson (2003a), the coherence of these peaks offers a strong argument for the theory of inflation.

The key work that established the utility of the anisotropy of CMB statistics to infer the lensing potential was the short paper by Hu (2001), which also contains the optimal estimator alluded to in §9.4. Hu and Okamoto (2002) followed up by pointing out that the polarization of the CMB also affords an opportunity to learn about the lensing potential, and indeed is likely to become more powerful in this regard. The first detection of CMB lensing was in cross-correlation (Smith et al., 2007) (this often happens with a new effect because the noise is smaller). The map shown in Fig. 9.4 comes from the Planck satellite mission (Ade et al., 2014); high-resolution ground-based experiments have produced higher signal-to-noise maps over smaller regions; one example of this is from the South Pole Telescope (van Engelen et al., 2012).

Exercises

9.1 Plot the intensity $I_\nu = (2h\nu^3/c^2)\left[e^{h\nu/k_B T} - 1\right]^{-1}$ as a function of frequency for a bosonic gas with a blackbody distribution and temperature $T = 2.75$K. Show that the intensity peaks at $\nu = 160$ GHz.

9.2 The temperature of the CMB scales as $a^{-1} = (1 + z)$. Find the value of the redshift at which the temperature was $13.6 eV/k_B$. Find the redshift when $k_B T = 1/4$ eV.

9.3 Determine the energy density of photons today by integrating the intensity over all frequency. Find the total number density of photons today by dividing the intensity by the photon energy $h\nu$ and then integrating over all frequency.

9.4 Using the results of Exercise (9.3), compare the number density of cosmic electrons (whether ionized or not) and photons in the CMB. Show that there are many more photons than electrons. To obtain the number density of electrons, divide the baryon density $\Omega_b \rho_{cr}$ by the energy per proton $m_p c^2$.

9.5 Show that if $f(x)$ is a real function of x, then its Fourier transform is complex but has the property that $\tilde{f}(k) = \tilde{f}^*(-k)$.

9.6 Compute the error on C_l in a bin in l-space of width $\Delta l = 100$ centered at $l = 1500$ in a CMB anisotropy experiment. Assume that experimental noise is negligible and the observation is of one square degree (this affects how many modes are measured); use the analog of Eq. (8.48). Compare with the error bar shown in Fig. 9.3.

9.7 Consider the product of two Fourier transforms $\tilde{f}(k) \times \tilde{g}(k)$. By writing each function in terms of its Fourier transform $f(x)$ and $g(x)$, show that this product is equal to a convolution in real space:

$$\int dx' f(x') g(x - x').$$

(9.27)

9.8 Carry out the angular part of the d^2l' integral in Eq. (9.26) and rewrite the integral as logarithmic in scale $\int (dl'/l') I(l')$; determine the integrand $I(l')$ in terms of the power spectrum of the deflection angle. Show that the relevant amplitude is that shown in Fig. 9.2.

Appendix A

Numbers

- Useful expression for Planck's constant

$$hc = 4.135667 \times 10^{-15} \, \text{eV sec}$$

- Boltzmann constant

$$k_B = 8.6173303 \times 10^{-5} \, \text{eV K}^{-1}$$

- Parsec

$$1 \, \text{pc} = 3.09 \times 10^{13} \text{km}$$
$$= 206264.81 \, \text{AU}$$

- Solar Mass

$$M_\odot = 1.989 \times 10^{33} \, \text{gm}$$

- Useful expressions for Newton's constant

$$MG_\odot = 4.302 \times 10^{-3} \, \text{pc} \, (\text{km/s})^2$$
$$MG_\odot/c^2 = 4.79 \times 10^{-14} \, \text{pc}$$

- 1 radian $= 57.30° = 3438$ arcminutes $= 2.063 \times 10^5$ arcseconds
- $M_{\text{Jupiter}} = 9.546 \times 10^{-4} \, M_\odot$
- $M_\oplus = 3.0 \times 10^{-6} M_\odot$
- $H_0 = 100h \, \text{km sec}^{-1} \, \text{Mpc}^{-1}$
- Critical density of the universe

$$\rho_{\text{cr}} \equiv \frac{3H_0^2}{8\pi G}$$
$$= 2.77 \times 10^{11} h^2 \, M_\odot \, \text{Mpc}^{-3}$$

Appendix B

Lensing Formulae

- Einstein Radius of point mass

$$\theta_E \equiv \sqrt{\frac{4MGD_{SL}}{D_S D_L c^2}} = 0.0028'' \left(\frac{\text{kpc}}{D_L}\right)^{1/2} \left(\frac{M}{M_\odot}\right)^{1/2} \left(\frac{D_{SL}}{D_S}\right)^{1/2} \quad \text{(B.1)}$$

- Critical Surface Density

$$\Sigma_{\text{cr}} = \frac{D_S c^2}{4\pi G D_{SL} D_L} \quad \text{(B.2)}$$

- Lens Equation

$$\vec{\beta} = \vec{\theta} - \vec{\alpha} \quad \text{(B.3)}$$

- Projected Potential

$$\Phi(\vec{\theta}) \equiv \frac{2}{D_S} \int_0^{D_S} dD_L \, \phi(x^i = D_L \theta^i, D_L; t = t_0 - D_L/c) \frac{D_{SL}}{D_L} \quad \text{(B.4)}$$

- Deflection Angle

$$\vec{\alpha} = \frac{\vec{\nabla}\Phi}{c^2} \quad \text{(B.5)}$$

- Distortion Tensor

$$\Psi_{ij} \equiv \frac{\partial \alpha_i}{\partial \theta_j} \equiv \frac{1}{c^2} \frac{\partial^2 \Phi}{\partial \theta_i \partial \theta_j}$$

$$= \begin{pmatrix} \kappa + \gamma_1 & \gamma_2 \\ \gamma_2 & \kappa - \gamma_1 \end{pmatrix} \quad \text{(B.6)}$$

- Two components of shear

$$\gamma_1 \equiv \frac{1}{2c^2} \left[\frac{\partial^2 \Phi}{\partial \theta_1^2} - \frac{\partial^2 \Phi}{\partial \theta_2^2}\right]$$

$$\gamma_2 \equiv \frac{1}{c^2} \frac{\partial^2 \Phi}{\partial \theta_1 \partial \theta_2} \tag{B.7}$$

- Convergence

$$\kappa \equiv \frac{1}{2c^2} \left[\frac{\partial^2 \Phi}{\partial \theta_1^2} + \frac{\partial^2 \Phi}{\partial \theta_2^2} \right]$$

$$= \frac{\Sigma}{\Sigma_{\mathrm{cr}}} \qquad \text{Thin lens} \tag{B.8}$$

- Surface Density

$$\Sigma(x, y) \equiv \int_{-\infty}^{\infty} dz \, \rho(x, y, z) \tag{B.9}$$

- Comoving Distance

$$\chi(z) = c \int_0^z \frac{dz'}{H(z')}$$

$$\to \frac{cz}{H_0} \qquad \text{Low} - z \, \text{limit} \tag{B.10}$$

- Lensing Distances

$$D_L \equiv \frac{\chi(z_L)}{1 + z_L}$$

$$D_S \equiv \frac{\chi(z_S)}{1 + z_S}$$

$$D_{SL} \equiv \frac{\chi(z_S) - \chi(z_L)}{1 + z_S} \tag{B.11}$$

References

Abbott, T., et al. 2005. The Dark Energy Survey. astro-ph/0510346.

Abbott, T., et al. 2016. Cosmology from cosmic shear with Dark Energy Survey science verification data. *Phys. Rev.*, **D94**(2), 022001.

Abell, G. O., Corwin, Jr., H. G., and Olowin, R. P. 1989. A catalog of rich clusters of galaxies. *Astrophysical Journal Supplement*, **70**(May), 1–138.

Ade, P. A. R., et al. 2014. Planck 2013 results. XVII. Gravitational lensing by large-scale structure. *Astron. Astrophys.*, **571**, A17.

Alcock, C., et al. 2000. The MACHO project: Microlensing results from 5.7 years of LMC observations. *Astrophys. J.*, **542**, 281–307.

Anderson, L., et al. 2014. The clustering of galaxies in the SDSS-III Baryon Oscillation Spectroscopic Survey: baryon acoustic oscillations in the Data Releases 10 and 11 Galaxy samples. *Mon. Not. Roy. Astron. Soc.*, **441**(1), 24–62.

Applegate, D. E., von der Linden, A., Kelly, P. L., Allen, M. T., Allen, S. W., Burchat, P. R., Burke, D. L., Ebeling, H., Mantz, A., and Morris, R. G. 2014. Weighing the Giants – III. Methods and measurements of accurate galaxy cluster weak-lensing masses. *Monthly Notices of the Royal Astronomical Society*, **439**(Mar.), 48–72.

Bartelmann, M., and Schneider, P. 2001. Weak gravitational lensing. *Phys. Rept.*, **340**, 291–472.

Becker, M. R., et al. 2016. Cosmic shear measurements with Dark Energy Survey science verification data. *Phys. Rev.*, **D94**(2), 022002.

Bekenstein, J. D. 2004. Relativistic gravitation theory for the MOND paradigm. *Phys. Rev.*, **D70**, 083509. [Erratum: Phys. Rev.D71,069901(2005)].

Bernstein, G. M., and Jarvis, M. 2002. Shapes and shears, stars and smears: optimal measurements for weak lensing. *Astronomical Journal*, **123**(Feb.), 583–618.

Bhattacharjee, Y., and Clery, D. 2013. A gallery of planet hunters. *Science*, **340**(6132), 566–569.

Blandford, R., and Narayan, R. 1986. Fermat's principle, caustics, and the classification of gravitational lens images. *Astrophysical Journal*, **310**(Nov.), 568–582.

Blandford, R. D., Saust, A. B., Brainerd, T. G., and Villumsen, J. V. 1991. The distortion of distant galaxy images by large-scale structure. *Monthly Notices of the Royal Astronomical Society*, **251**(Aug.), 600–627.

Bridle, S. L., Kneib, J.-P., Bardeau, S., and Gull, S. F. 2002 (Mar.). Bayesian galaxy shape estimation. Pages 38–46 of: Natarajan, P. (ed), *The Shapes of Galaxies and their Dark Halos*.

Caldwell, R. R., and Kamionkowski, M. 2009. The physics of cosmic acceleration. *Ann. Rev. Nucl. Part. Sci.*, **59**, 397–429.

Carroll, B. W., and Ostlie, D. A. 2006. *An Introduction to Modern Astrophysics and Cosmology*. Addison-Wesley.

Carroll, S. M. 2004. *Spacetime and Geometry: An Introduction to General Relativity*. Addison-Wesley.

Catelan, P., Kamionkowski, M., and Blandford, R. D. 2001. Intrinsic and extrinsic galaxy alignment. *Monthly Notices of the Royal Astronomical Society*, **320**, L7–L13.

Coles, P., and Lucchin, F. 1995. *Cosmology: The Origin and Evolution of Cosmic Structure*. John Wiley.

Dalal, N., and Kochanek, C. S. 2002. Direct detection of CDM substructure. *Astrophys. J.*, **572**, 25–33.

de Lapparent, V., Geller, M. J., and Huchra, J. P. 1986. A slice of the universe. *Astrophysical Journal Letters*, **302**(Mar.), L1–L5.

Dicke, R. H., Peebles, P. J. E., Roll, P. G., and Wilkinson, D. T. 1965. Cosmic black-body radiation. *Astrophysical Journal*, **142**(July), 414–419.

Dodelson, S. 2003a. Coherent phase argument for inflation. *AIP Conf. Proc.*, **689**, 184–196.

Dodelson, S. 2003b. *Modern Cosmology*. Amsterdam: Academic Press.

Dodelson, S. 2011. The real problem with MOND. *Int. J. Mod. Phys.*, **D20**, 2749–2753.

Dwek, E., Arendt, R. G., Hauser, M. G., Kelsall, T., Lisse, C. M., Moseley, S. H., Silverberg, R. F., Sodroski, T. J., and Weiland, J. L. 1995. Morphology, near-infrared luminosity, and mass of the Galactic bulge from COBE DIRBE observations. *Astrophysical Journal*, **445**(June), 716–730.

Dyson, F. W., Eddington, A. S., and Davidson, C. 1920. A determination of the deflection of light by the sun's gravitational field, from observations made at the total eclipse of May 29, 1919. *Philosophical Transactions of the Royal Society of London Series A*, **220**, 291–333.

Ebeling, H., Edge, A. C., Mantz, A., Barrett, E., Henry, J. P., Ma, C. J., and van Speybroeck, L. 2010. The X-ray brightest clusters of galaxies from the Massive Cluster Survey. *Monthly Notices of the Royal Astronomical Society*, **407**(Sept.), 83–93.

Einstein, A. 1936. Lens-like action of a star by the deviation of light in the gravitational field. *Science*, **84**(Dec.), 506–507.

Fadely, R., Keeton, C. R., Nakajima, R., and Bernstein, G. M. 2010. Improved constraints on the gravitational lens Q0957+561. II. Strong lensing. *Astrophysical Journal*, **711**(Mar.), 246–267.

Fassnacht, C. D., Blandford, R. D., Cohen, J. G., Matthews, K., Pearson, T. J., Readhead, A. C. S., Womble, D. S., Myers, S. T., Browne, I. W. A., Jackson, N. J., Marlow, D. R., Wilkinson, P. N., Koopmans, L. V. E., de Bruyn, A. G., Schilizzi, R. T., Bremer, M., and Miley, G. 1999. B2045+265: A new four-image gravitational lens from CLASS. *Astronomical Journal*, **117**(Feb.), 658–670.

Foreman-Mackey, D., Hogg, D. W., Lang, D., and Goodman, J. 2013. emcee: The MCMC Hammer. *Publications of the Astronomical Society of Pacific*, **125**(Mar.), 306–312.

Franx, M., Illingworth, G. D., Kelson, D. D., van Dokkum, P. G., and Tran, K.-V. 1997. A pair of lensed galaxies at z = 4.92 in the field of CL 1358+62. *Astrophysical Journal Letters*, **486**(Sept.), L75–L78.

Frieman, J., Turner, M., and Huterer, D. 2008. Dark energy and the accelerating universe. *Ann. Rev. Astron. Astrophys.*, **46**, 385–432.

Gates, E. 2009. *Einstein's Telescope: The Hunt for Dark Matter and Dark Energy in the Universe*. W.W. Norton & Company.

Gaudi, B. Scott. 2010. Exoplanetary microlensing. *arXiv:1002.0332*.

Gould, A., and Loeb, A. 1992. Discovering planetary systems through gravitational microlenses. *Astrophysical Journal*, **396**(Sept.), 104–114.

Griest, K. 1991. Galactic microlensing as a method of detecting massive compact halo objects. *Astrophysical Journal*, **366**(Jan.), 412–421.

Han, C., and Gould, A. P. 2003. Stellar contribution to the Galactic bulge microlensing optical depth. *Astrophys.J.*, **592**, 172–175.

Hartle, J. B. 2003. *An introduction to Einstein's General Relativity*. Addison-Wesley.

Heymans, C., et al. 2013. CFHTLenS tomographic weak lensing cosmological parameter constraints: Mitigating the impact of intrinsic galaxy alignments. *Mon. Not. Roy. Astron. Soc.*, **432**, 2433.

Hirata, C. M., and Seljak, U. 2004. Intrinsic alignment-lensing interference as a contaminant of cosmic shear. *Physical Review D*, **70**(6), 063526.

Hirata, C. M., and Seljak, U. 2003. Shear calibration biases in weak lensing surveys. *Mon. Not. Roy. Astron. Soc.*, **343**, 459–480.

Hoekstra, H., Franx, M., Kuijken, K., and Squires, G. 1998. Weak lensing analysis of CL 1358+62 using Hubble Space Telescope observations. *Astrophysical Journal*, **504**(Sept.), 636–660.

Hoekstra, H., and Jain, B. 2008. Weak gravitational lensing and its cosmological applications. *Ann. Rev. Nucl. Part. Sci.*, **58**, 99–123.

Hogg, D. W. 1999. Distance measures in cosmology. *astro-ph/9905116*.

Hu, W. 1999. Power spectrum tomography with weak lensing. *Astrophys. J.*, **522**, L21–L24.

Hu, W. 2001. Mapping the dark matter through the cmb damping tail. *Astrophys. J.*, **557**, L79–L83.

Hu, W., and Okamoto, T. 2002. Mass reconstruction with cmb polarization. *Astrophys. J.*, **574**, 566–574.

Huff, E. M., and Graves, G. J. 2014. Magnificent magnification: exploiting the other half of the lensing signal. *Astrophys. J.*, **780**, L16.

Jarvis, M., et al. 2015. The DES Science Verification Weak Lensing Shear Catalogs.

Kaiser, N. 1992. Weak gravitational lensing of distant galaxies. *Astrophysical Journal*, **388**(Apr.), 272–286.

Kaiser, N., Squires, G., and Broadhurst, T. 1995. A method for weak lensing observations. *Astrophysical Journal*, **449**(Aug.), 460.

Kaiser, N., and Squires, G. 1993. Mapping the dark matter with weak gravitational lensing. *Astrophys. J.*, **404**, 441–450.

Keeton, C. R., Falco, E. E., Impey, C. D., Kochanek, C. S., Lehár, J., McLeod, B. A., Rix, H.-W., Muñoz, J. A., and Peng, C. Y. 2000. The host galaxy of the lensed quasar Q0957+561. *Astrophysical Journal*, **542**(Oct.), 74–93.

Kelly, P. L., et al. 2015. Multiple images of a highly magnified supernova formed by an early-type cluster galaxy lens. *Science*, **347**, 1123.

Kennefick, D. 2007. Not only because of theory: Dyson, Eddington and the competing myths of the 1919 eclipse expedition. *arXiv/physics.hist-ph/0709.0685*.

Kilbinger, M., Fu, L., Heymans, C., Simpson, F., Benjamin, J., Erben, T., Harnois-Déraps, J., Hoekstra, H., Hildebrandt, H., Kitching, T. D., Mellier, Y., Miller, L., Van Waerbeke, L., Benabed, K., Bonnett, C., Coupon, J., Hudson, M. J., Kuijken, K., Rowe, B., Schrabback, T., Semboloni, E., Vafaei, S., and Velander, M. 2013. CFHTLenS: combined probe cosmological model comparison using 2D weak gravitational lensing. *Monthly Notices of the Royal Astronomical Society*, **430**(Apr.), 2200–2220.

Kuijken, K. 1999. Weak weak lensing: correcting weak shear measurements accurately for psf anisotropy. *Astron. Astrophys.*, **352**, 355.

Lawrence, A., Rowan-Robinson, M., Ellis, R. S., Frenk, C. S., Efstathiou, G., Kaiser, N., Saunders, W., Parry, I. R., Xiaoyang, X., and Crawford, J. 1999. The QDOT all-sky IRAS galaxy redshift survey. *Monthly Notices of the Royal Astronomical Society*, **308**(Oct.), 897–927.

Liddle, A. 2003. *An Introduction to Modern Cosmology, Second Edition*. John Wiley & Sons.

Limber, D. N. 1953. The analysis of counts of the extragalactic nebulae in terms of a fluctuating density field. *Astrophysical Journal*, **117**(Jan.), 134.

Lin, H., Dodelson, S., Seo, H.-J., Soares-Santos, M., Annis, J., Hao, J., Johnston, D., Kubo, J. M., Reis, R. R. R., and Simet, M. 2012. The SDSS co-add: cosmic shear measurement. *Astrophysical Journal*, **761**(Dec.), 15.

Linder, E. V. 2011. Lensing time delays and cosmological complementarity. *Physical Review D*, **84**(12), 123529.

LoVerde, M., and Afshordi, N. 2008. Extended limber approximation. *Phys. Rev.*, **D78**, 123506.

Mandelbaum, R., et al. 2014. The third gravitational lensing accuracy testing (GREAT3) challenge handbook. *Astrophys. J. Suppl.*, **212**, 5.

Martinez, V. J., and Coles, P. 1994. Correlations and scaling in the QDOT redshift survey. *Astrophysical Journal*, **437**(Dec.), 550–555.

McKay, T. A., Sheldon, E. S., Racusin, J., Fischer, P., Seljak, U., Stebbins, A., Johnston, D., Frieman, J. A., Bahcall, N., Brinkmann, J., Csabai, I., Fukugita, M., Hennessy, G. S., Ivezic, Z., Lamb, D. Q., Loveday, J., Lupton, R. H., Munn, J. A., Nichol, R. C., Pier, J. R., and York, D. G. 2001. Galaxy mass and luminosity scaling laws determined by weak gravitational lensing. *astro-ph/0108013*.

Melchior, P., Viola, M., Schafer, B. M., and Bartelmann, M. 2011. Weak gravitational lensing with DEIMOS. *Mon. Not. Roy. Astron. Soc.*, **412**, 1552.

Milgrom, M. 1983. A modification of the Newtonian dynamics as a possible alternative to the hidden mass hypothesis. *Astrophysical Journal*, **270**(July), 365–370.

Minniti, D., and Zoccali, M. 2008. The galactic bulge: a review. *IAU Symp.*, **245**, 323.

Miralda-Escude, J. 1991. The correlation function of galaxy ellipticities produced by gravitational lensing. *Astrophysical Journal*, **380**(Oct.), 1–8.

Mollerach, S., and Roulet, E. 2002. *Gravitational Lensing and Microlensing*. World Scientific Pub Co Inc (January 31, 2002).

Navarro, J. F., Frenk, C. S., and White, S. D. M. 1997. A Universal Density Profile from Hierarchical Clustering. *Astrophysical Journal*, **490**(Dec.), 493–508.

Oguri, M., and Marshall, P. J. 2010. Gravitationally lensed quasars and supernovae in future wide-field optical imaging surveys. *Monthly Notices of the Royal Astronomical Society*, **405**(July), 2579–2593.

Paczynski, B. 1986. Gravitational microlensing by the galactic halo. *Astrophys.J.*, **304**, 1–5.

Peacock, J. A. 1999. *Cosmological Physics; Rev. Version*. Cambridge: Cambridge University Press.

Peebles, P. J. E., Page, Jr., L. A., and Partridge, R. B. 2009. *Finding the Big Bang*. Cambridge: Cambridge University Press.

Penzias, A. A., and Wilson, R. W. 1965. A measurement of excess antenna temperature at 4080 Mc/s. *Astrophysical Journal*, **142**(July), 419–421.

Popowski, P., Griest, K., Thomas, C. L., Cook, K. H., Bennett, D. P., Becker, A. C., Alves, D. R., Minniti, D., Drake, A. J., Alcock, C., Allsman, R. A., Axelrod, T. S., Freeman, K. C., Geha, M., Lehner, M. J., Marshall, S. L., Nelson, C. A., Peterson, B. A., Quinn,

P. J., Stubbs, C. W., Sutherland, W., Vandehei, T., Welch, D., and MACHO Collaboration. 2005. Microlensing optical depth toward the galactic bulge using clump giants from the MACHO Survey. *Astrophysical Journal*, **631**(Oct.), 879–905.

Refregier, A. 2003. Shapelets: I. a method for image analysis. *Mon. Not. Roy. Astron. Soc.*, **338**, 35.

Rhodes, J., Allen, S., Benson, B. A., Chang, T., de Putter, R., et al. 2015. Exploiting cross correlations and joint analyses. *Astropart. Phys.*, **63**, 42–54.

Ruel, J., Bazin, G., Bayliss, M., Brodwin, M., Foley, R. J., et al. 2014. Optical spectroscopy and velocity dispersions of galaxy clusters from the SPT-SZ Survey. *Astrophysical Journal*, **792**, 45.

Ryden, B. 2003. *Introduction to Cosmology*. Addison-Wesley.

Sanders, R. H., and McGaugh, S. S. 2002. Modified Newtonian dynamics as an alternative to dark matter. *Annual Review of Astronomy and Astrophysics*, **40**, 263–317.

Schmidt, F., Leauthaud, A., Massey, R., Rhodes, J., George, M. R., Koekemoer, A. M., Finoguenov, A., and Tanaka, M. 2012. A detection of weak-lensing magnification using galaxy sizes and magnitudes. *Astrophysical Journal Letters*, **744**(Jan.), L22.

Schneider, P. 2006. *Extragalactic Astronomy and Cosmology*. Springer.

Schneider, P., Ehlers, J., and Falco, E. E. 1992. *Gravitational Lenses*.

Schneider, P., Meylan, G., Kochanek, C., Jetzer, P., North, P., and Wambsganss, J. 2006. *Gravitational Lensing: Strong, Weak and Micro: Saas-Fee Advanced Course 33*. Saas-Fee Advanced Course. Springer.

Schutz, B. F. 1985. *A First Course in General Relativity*. Cambridge: Cambridge University Press.

Serjeant, S. 2010. *Observational Cosmology*. Cambridge: Cambridge University Press.

Shapiro, I. I. 1964. Fourth test of general relativity. *Phys. Rev. Lett.*, **13**, 789–791.

Shapiro, I. I., Pettengill, Gordon H., Ash, M. E., Stone, M. L., Smith, W. B., et al. 1968. Fourth test of general relativity: preliminary results. *Phys. Rev. Lett.*, **20**, 1265–1269.

Silver, N. 2012. *The Signal and the Noise: Why So Many Predictions Fail–But Some Don't*. Penguin Publishing Group.

Skowron, J., Shin, I.-G., Udalski, A., Han, C., Sumi, T., Shvartzvald, Y., Gould, A., Dominis Prester, D., Street, R. A., Jørgensen, U. G., Bennett, D. P., Bozza, V., Szymański, M. K., Kubiak, M., Pietrzyński, G., Soszyński, I., Poleski, R., Kozłowski, S., Pietrukowicz, P., Ulaczyk, K., Wyrzykowski, Ł., OGLE Collaboration, Abe, F., Bhattacharya, A., Bond, I. A., Botzler, C. S., Freeman, M., Fukui, A., Fukunaga, D., Itow, Y., Ling, C. H., Koshimoto, N., Masuda, K., Matsubara, Y., Muraki, Y., Namba, S., Ohnishi, K., Philpott, L. C., Rattenbury, N., Saito, T., Sullivan, D. J., Suzuki, D., Tristram, P. J., Yock, P. C. M., MOA Collaboration, Maoz, D., Kaspi, S., Friedmann, M., Wise Group, Almeida, L. A., Batista, V., Christie, G., Choi, J.-Y., DePoy, D. L., Gaudi, B. S., Henderson, C., Hwang, K.-H., Jablonski, F., Jung, Y. K., Lee, C.-U., McCormick, J., Natusch, T., Ngan, H., Park, H., Pogge, R. W., Yee, J. C., μFUN Collaboration, Albrow, M. D., Bachelet, E., Beaulieu, J.-P., Brillant, S., Caldwell, J. A. R., Cassan, A., Cole, A., Corrales, E., Coutures, C., Dieters, S., Donatowicz, J., Fouqué, P., Greenhill, J., Kains, N., Kane, S. R., Kubas, D., Marquette, J.-B., Martin, R., Menzies, J., Pollard, K. R., Ranc, C., Sahu, K. C., Wambsganss, J., Williams, A., Wouters, D., PLANET Collaboration, Tsapras, Y., Bramich, D. M., Horne, K., Hundertmark, M., Snodgrass, C., Steele, I. A., RoboNet Collaboration, Alsubai, K. A., Browne, P., Burgdorf, M. J., Calchi Novati, S., Dodds, P., Dominik, M., Dreizler, S., Fang, X.-S., Gu, C.-H., Hardis, S., Harpsøe, K., Hessman, F. V., Hinse, T. C., Hornstrup, A., Jessen-Hansen, J., Kerins, E., Liebig, C., Lund, M., Lundkvist, M., Mancini, L., Mathiasen, M., Penny, M. T., Rahvar, S., Ricci, D., Scarpetta, G., Skottfelt, J., Southworth, J.,

Surdej, J., Tregloan-Reed, J., Wertz, O., and MiNDSTEp Consortium. 2015. OGLE-2011-BLG-0265Lb: A Jovian microlensing planet orbiting an M dwarf. *Astrophysical Journal*, **804**(May), 33.

Smith, K. M., Zahn, O., and Dore, O. 2007. Detection of gravitational lensing in the Cosmic Microwave Background. *Phys. Rev.*, **D76**, 043510.

Sofue, Y. 2012. Grand rotation curve and dark matter halo in the Milky Way Galaxy. *Publications of the Astronomical Society of Japan*, **64**(Aug.), 75.

Spergel, D., Gehrels, N., Baltay, C., Bennett, D., Breckinridge, J., Donahue, M., Dressler, A., Gaudi, B. S., Greene, T., Guyon, O., Hirata, C., Kalirai, J., Kasdin, N. J., Macintosh, B., Moos, W., Perlmutter, S., Postman, M., Rauscher, B., Rhodes, J., Wang, Y., Weinberg, D., Benford, D., Hudson, M., Jeong, W.-S., Mellier, Y., Traub, W., Yamada, T., Capak, P., Colbert, J., Masters, D., Penny, M., Savransky, D., Stern, D., Zimmerman, N., Barry, R., Bartusek, L., Carpenter, K., Cheng, E., Content, D., Dekens, F., Demers, R., Grady, K., Jackson, C., Kuan, G., Kruk, J., Melton, M., Nemati, B., Parvin, B., Poberezhskiy, I., Peddie, C., Ruffa, J., Wallace, J. K., Whipple, A., Wollack, E., and Zhao, F. 2015. Wide-field infrared survey telescope-Astrophysics Focused Telescope Assets WFIRST-AFTA 2015 Report. *ArXiv e-prints, 1503.03757*.

Squires, G., Kaiser, N., Babul, A., Fahlman, G., Woods, D., Neumann, D. M., and Boehringer, H. 1996. The dark matter, gas, and galaxy distributions in Abell 2218: A weak gravitational lensing and X-ray analysis. *Astrophysical Journal*, **461**(Apr.), 572.

Suyu, S. H., Auger, M. W., Hilbert, S., Marshall, P. J., Tewes, M., Treu, T., Fassnacht, C. D., Koopmans, L. V. E., Sluse, D., Blandford, R. D., Courbin, F., and Meylan, G. 2013. Two accurate time-delay distances from strong lensing: implications for cosmology. *Astrophysical Journal*, **766**(Apr.), 70.

Treu, T., et al. 2016. Refsdal meets Popper: Comparing predictions of the re-appearance of the multiply imaged supernova behind MACSJ1149.5+2223. *Astrophys. J.*, **817**(1), 60.

Troxel, M. A., and Ishak, M. 2015. The intrinsic alignment of galaxies and its impact on weak gravitational lensing in an era of precision cosmology. *Physics Reports*, **558**(Feb.), 1–59.

Tyson, J. A., Wenk, R. A., and Valdes, F. 1990. Detection of systematic gravitational lens galaxy image alignments – Mapping dark matter in galaxy clusters. *Astrophysical Journal Letters*, **349**(Jan.), L1–L4.

van Engelen, A., et al. 2012. A measurement of gravitational lensing of the microwave background using South Pole Telescope data. *Astrophys. J.*, **756**, 142.

Vikram, V., et al. 2015. Wide-field lensing mass maps from Dark Energy Survey science verification data: Methodology and detailed analysis. *Phys. Rev.*, **D92**(2), 022006.

Walsh, D., Carswell, R. F., and Weymann, R. J. 1979. 0957 + 561 A, B – Twin quasistellar objects or gravitational lens. *Nature*, **279**(May), 381–384.

Wambsganss, J. 1998. Gravitational lensing in astronomy. *Living Rev. Rel.*, **1**, 12.

Weinberg, D., Bard, D., Dawson, K., Dore, O., Frieman, J., Gebhardt, K., Levi, M., and Rhodes, J. 2013. Facilities for dark energy investigations. *arXiv/astro-ph.CO/1309.5380*.

Weinberg, S. 2008. *Cosmology*. Oxford: Oxford University Press.

Will, C. 1993. *Theory and Experiment in Gravitational Physics*. Cambridge, England; New York, NY, USA: Cambridge University Press.

Will, C. M. 2014. The confrontation between general relativity and experiment. *Living Rev.Rel.*, **17**, 4.

Wright, C., and Brainerd, T. G. 1999. Gravitational lensing by NFW Halos. *astro-ph/9908213*.

Wright, J. T., and Gaudi, B. S. 2013. Exoplanet detection methods. Page 489 of: Oswalt, T. D., French, L. M., and Kalas, P. (eds), *Planets, Stars and Stellar Systems. Volume 3: Solar and Stellar Planetary Systems.*

Zentner, A. R. 2007. The excursion set theory of halo mass functions, halo clustering, and halo growth. *Int. J. Mod. Phys.*, **D16**, 763–816.

Zitrin, A., Zheng, W., Broadhurst, T., Moustakas, J., Lam, D., Shu, X., Huang, X., Diego, J. M., Ford, H., Lim, J., Bauer, F. E., Infante, L., Kelson, D. D., and Molino, A. 2014. A geometrically supported z ~ 10 candidate multiply imaged by the Hubble Frontier Fields Cluster A2744. *Astrophysical Journal Letters*, **793**(Sept.), L12.

Zwicky, F. 1937. Nebulae as gravitational lenses. *Phys. Rev.*, **51**(Feb), 290–290.

Index

ΛCDM, 166, 184, 192
σ_8, 190

Abell catalog, 145
aberration, 119
accelerating universe, 145, 165, 192
acoustic oscillations
  damping, 198
  fundamental mode, 198
  higher harmonics, 198
Active Galactic Nucleus, AGN, 57
additive bias, 126
affine parameter, 23
anisotropy spectrum, 13
apparent magnitude, 75
arcsecond, 1
astronomical unit, AU, 98, 104
atter power spectrum, 178

B-mode
  of shear, 135
baryon density, Ω_b, 165
Baryon Oscillation Spectroscopic Survey (BOSS),
    170
baryons, 165
Bayes' Theorem, 160
black holes, 89
Boltzmann constant, 196, 210
Born approximation, 31
boson, 196
Bremsstrahlung, 146

Canada–France–Hawaii Telescope (CFHT), 179, 189
Cassini spacecraft, 28
caustics, 50
  as curves of infinite magnification, 66
  asymmetric lens, 80
  isothermal sphere, 52, 72

center of mass
  of star and planet, 95
Chandra satellite, 146
Christoffel symbol, 22, 141
circular PSF
  effect on ellipticity, 120
cluster, *see* galaxy cluster
clustering of matter
  using tomography, 188
CMB, *see* Cosmic Microwave Background
Cold Dark Matter, CDM, 166
comoving coordinates, 163
comoving distance, 212
complex conjugate, 203
Compton scattering, 197
conditional probability, 160
conjugate variables, 155
  l and θ, 200
convergence, 11, 34, 64, 154, 163, 212
  as delineator of strong and weak lensing, 35
  as dilation, 65
  as element of distortion tensor, 64
  effect on acoustic peaks, 200–201
  isothermal sphere, 69
  relation to Φ, 37
  relation to Φ in Fourier space, 156, 163
  spherically symmetric, 42–44
convolution, 203, 209
core angular radius, 46
core radius, 44
correlation function, 167, 175
Cosmic Microwave Background, CMB, 3, 12, 165,
    196–199
  lensing, 196
  temperature of, 14, 204
cosmic shear, 11, 135, 162
  E-B decomposition, 173
  B-modes, 175
  E-modes, 173

220